Self-Consolidating Concrete

Self-consolidating concrete (SCC) is an innovative material used successfully throughout the world. It is a highly flowable, non-segregating concrete that can spread into place, fill the formwork, and encapsulate the reinforcement without any mechanical consolidation, improving the overall efficiency of a concrete construction project. SCC mixtures are highly fluid, yet their flowing properties can be adapted for a range of applications and allow practitioners to select and determine levels of filling ability, passing ability, and stability.

Self-Consolidating Concrete: Applying What We Know discusses all aspects of SCC, including:

- Benefits and limitations
- Raw material components
- Mixture proportions
- Production and quality control
- Placement and curing.

Presenting a basis for consistently producing and applying SCC in a regular production environment, this book is written from the perspective of the concrete practitioner. It uses descriptions and case studies throughout, as examples of specific types of applications, to identify where the practitioner needs to focus attention.

This book bridges the gap between research and practice. It links science with practical application, describing a number of projects and types of applications where SCC has been used successfully. It will be useful for new practitioners as well as for those already using SCC.

Joseph A. Daczko is a concrete technologist with the admixture systems business of BASF. He was the first chairman of the American Concrete Institute's committee on Self-Consolidating Concrete and was presented with ACI's Delmar Bloem Award for leadership of this committee.

Self-Consolidating Concrete

Applying What We Know

Joseph A. Daczko

CRC Press
Taylor & Francis Group
Boca Raton London New York

CRC Press is an imprint of the
Taylor & Francis Group, an **informa** business
A SPON PRESS BOOK

CRC Press
Taylor & Francis Group
6000 Broken Sound Parkway NW, Suite 300
Boca Raton, FL 33487-2742

First issued in paperback 2019

© 2012 Joseph A. Daczko
CRC Press is an imprint of Taylor & Francis Group, an Informa business

The right of Joseph A. Daczko to be identified as author of this work has been asserted by him in accordance with sections 77 and 78 of the Copyright, Designs and Patents Act 1988.

ISBN-13: 978-0-415-59064-8 (hbk)
ISBN-13: 978-0-367-86492-7 (pbk)

British Library Cataloguing in Publication Data
A catalogue record for this book is available from the British Library

Library of Congress Cataloging in Publication Data
Daczko, Joseph A.
   Self-consolidating concrete : applying what we know / Joseph A. Daczko.
      p. cm.
   Includes bibliographical references and index.
   1. Self-consolidating concrete. 2. Concrete—Mixing. I. Title.
   TA442.5.D33 2012
   620.1′36—dc23   2011033602

Typeset in Sabon
by Keystroke, Station Road, Codsall, Wolverhampton

Visit the Taylor & Francis Web site at
http://www.taylorandfrancis.com

and the CRC Press Web site at
http://www.crcpress.com

This book is dedicated to "the past"—
In loving memory of Mary Virginia Daczko
and to "the future"—To Angela, Nathan, and Joey Daczko

Contents

Preface

When first confronted with the prospect of undertaking a task that will require a significant investment of one's time and energy, it is quite natural to have conflicting emotions. Such was the case when the idea of writing this book was first presented to me. In the "concrete world," self-consolidating/self-compacting concrete (SCC) has been the topic of numerous excellent publications, papers, and presentations over the last two decades. So many interesting and very insightful discussions have taken place around this technology (and I am sure there are more to come). With this understanding of having so much work already being done, and knowing that one book had already been published on this topic, I had to ask myself: "How can anyone really write anything knew or useful that hasn't already been said?" In my mind, this was the question that needed answering prior to moving forward with this project.

Over time, however, through informal discussions and conversations with numerous industry professionals and reviewing much of what had already been published, a concept began to materialize. Another book compiling information on the various material properties of SCC (although that is important and to a certain degree included) was not needed. What was needed was a book discussing the practical application of SCC—practical in the sense of materials selection and control, production, quality control, placement, and finishing. This is where I believe the current need is with respect to SCC technology.

All new innovations, whether in the world of concrete construction or in consumer electronics, go through a diffusion or acceptance process, and SCC is no different. Although the overall phases of the acceptance process may be similar across industries, the rate of diffusion or acceptance certainly differs. By comparison, the diffusion rate within the concrete construction industry versus other industries could be considered relatively slow. Of particular interest, however, is that the acceptance rate of SCC among segments of the concrete industry is quite different. For example, the precast/prestressed industry far surpasses the cast-in-place concrete segment in the percent of SCC used. The question is why? Why are cast-in-place projects not applying it at the same rate as precast/prestressed producers? Can what has been learned in one area be transferred to the other? What does it take to effectively implement SCC? What about those who have tried SCC before and had a negative experience?

From this perspective, many questions can be asked when determining why one is or is not using SCC. The reality is that there cannot be a single answer to any of these questions. The circumstances surrounding each plant and each project differ. Each person, team, or organization must answer the questions for themselves and work through the process to form their own conclusions. It is from the perspective of helping those involved to answer these practical questions for themselves that this book is written. There is certainly discussion on a variety of technical topics, but always with an eye toward application. It is my hope that this book will provide a level of insight or present new ideas that will allow a person or organization to answer these questions for themselves and for their particular circumstance and make the best possible use of this exciting technology.

Joseph A. Daczko
May 2011

Acknowledgments

There are many people who have in one way or another impacted the writing of this book. First and foremost, however, I must thank God for allowing me these many opportunities and providing me the Grace and courage to become involved. Second, I must thank my wife Angela and my two sons, Nathan and Joey. Your support of this work and of me can never be replaced or forgotten. Thank you.

I also thank BASF for giving me permission to write this book as well as to use much of the data we have generated over the last 15 years. What a great group of people! Special thanks are given to Emmanuel Attiogbe and the late Will Secre, who allowed me the time and freedom to explore the many areas of SCC technology—without their trust, this book would never have happened. Additional thanks are due to a number of people who reviewed and commented on either parts of the book or its entirety, including Peter Billberg, Daniel Constantiner, Tate Coverdale, John Luciano, Kevin MacDonald, Lloyd Keller, and Viswanath Mahadevan. And finally to Donald S. Daczko, engineer and contractor—thanks for reviewing the book Dad.

Chapter 1

Introduction, History, and Acceptance of SCC

Introduction

Self-consolidating or self-compacting concrete, often abbreviated SCC, may be one of the most significant concrete technology developments in many years. Because of its characteristic fresh properties, it has the potential to dramatically alter and improve the future of concrete placement and construction processes. In 1999, at the first RILEM symposium on SCC, held in Stockholm, Sweden, Ake Skarendahl suggested that with equipment advances, the use of SCC could allow automation of certain segments of the industry such as precast concrete production.[1] Although further equipment developments, particularly in the area of production control equipment, are still necessary, to some degree this evolution has begun. New precast production plants, in Europe and the United States, have been designed around the use of SCC.[2,3] As will be seen in later chapters of this book, the use of SCC may lead to labor and construction time savings, and result in architectural finishes that may not have been possible with conventional concrete. In addition to influencing project results and spurring developments around manufacturing plant design, SCC has advanced interest in other areas of materials science. For example, rheology, defined as the science of the deformation and flow of matter,[4] has become a much more talked about area of interest. Although this science was applied to concrete before the development of SCC,[5] the rheological properties of concrete were not widely discussed among material suppliers, concrete producers and contractors. Concrete rheology was relegated to mostly academic and research discussions. With the defining characteristics of SCC being its fresh, flow properties, however, SCC has become the type of concrete to which the principles of concrete rheology can be applied. The growing interest in SCC, and a practical understanding of its flow properties, have advanced the practical user's interest in concrete rheology and its application to conventional slump concrete.

Over the last decade, interest in SCC has continued to grow significantly; however, some hesitancy to fully embrace this technology still remains. This is likely due to the natural human resistance to change and/or a lack of understanding of the significant benefits that arise from the use of SCC. This book is written with the intention of bridging the gap between science and practice by providing a practical viewpoint relative to the application of SCC. It is the author's objective to make the concrete practitioner more comfortable with, and proficient in, the use of SCC.

Definitions of SCC

Throughout this book, SCC will be discussed from a number of viewpoints, and so it is important to start with a basic understanding of what SCC is and how it is different from conventional concrete. Multiple definitions exist from individuals and industry associations across Asia, Europe, and North America, three of which are presented here:

1. A concrete that "can be compacted into every corner of a formwork, purely by means of its own weight and without the need for vibrating compaction."[6]
2. Self-compacting concrete (SCC) is an innovative concrete that does not require vibration for placing and compaction. It is able to flow under its own weight, completely filling formwork and achieving full compaction, even in the presence of congested reinforcement. The hardened concrete is dense, homogeneous and has the same engineering properties and durability as traditional vibrated concrete.[7]
3. Self-consolidating concrete (SCC) is highly flowable, non-segregating concrete that can spread into place, fill the formwork, and encapsulate the reinforcement without any mechanical consolidation.[8]

SCC is clearly identified by its fresh properties, which will be reviewed in Chapter 3. But how are these properties different from those of conventional concrete? What about SCC's hardened properties, and the materials used to produce SCC? The easiest way to answer these questions is to compare SCC with conventional slump concrete, referred to from this point forward as conventional concrete.

How Does SCC Compare with Conventional Concrete?

Constituent Materials

SCC is composed of Portland cement, fine aggregate, coarse aggregate, water, chemical admixtures, and typically supplementary cementitious materials such as fly ash, slag, silica fume, and metakaolin. In some cases, mineral fillers such as limestone powder or very fine sands are used to increase the mixture's powder or fine material content. Aside from mineral fillers, the materials used to produce SCC are the same as those common to the production of conventional concrete. Because common materials are used, evaluating the performance of SCC relative to conventional concrete either in a laboratory setting or in full-scale production can be a fairly simple process. A more detailed discussion on material properties and measurements are presented in Chapters 5 and 11.

Mixture Proportions

Many documents present SCC mixture proportions, relative to conventional concrete, as typically having a higher cementitious or powder content, and containing a lower per unit volume of coarse aggregate.[7,9] However, when using relative terms such as "higher" or "lower" to describe SCC proportions, it is important to clearly understand the reference to which SCC is being compared. For example, the relative comparison of SCC mixture proportions to those of a residential slab or a prestressed bridge girder will be quite different.

Table 1.1 provides example mixture proportions for two conventional concrete mixtures and two SCC mixtures. Notice first that not all SCC mixtures are proportioned the same, as can be seen by comparing Mixtures 1 and 2 in Table 1.1. Similar to the way that hardened property requirements, such as compressive strength, influence how a mixture is proportioned, a range of SCC fresh property performance levels exist and can influence how SCC mixtures are proportioned, and therefore SCC mixture proportions will vary from application to application. The third and fourth mixtures presented in Table 1.1 are examples of a concrete slab mixture and a high-strength concrete mixture taken directly from ACI 302, "Guide for Concrete Floor and Slab Construction," and ACI 363.2R, "Guide to Quality Control and Testing of High Strength Concrete," respectively.[10,11] Comparing these mixture proportions with the example SCC proportions presented, one can conclude that SCC proportions tend to be more similar to those of high-strength concrete. In fact, ACI 237 states that SCC is a high-performance concrete mixture in the fresh state.[8] Remember, however, that the target properties of SCC in the fresh state can be higher or lower depending upon the application. Therefore, when making comparisons between SCC and conventional concrete mixture proportions, two things should be noted: the target property level of the SCC mixture being evaluated and the target property level of the mixture to which SCC is being compared. As the target properties of the mixture with which SCC is being compared decrease, and/or as the target properties of the SCC mixture increase, the differences in mixture proportions will tend to increase. SCC mixture proportions are reviewed and discussed more fully in Chapter 8.

Table 1.1 Example Constituent Material Proportions of Conventional Concrete and SCC

	1	*2*	*3*	*4*
	SCC 1[8]	SCC 2[8]	Concrete Slab[10]	High-Strength[11]
Cement (kg/m³)	309	445	285	451
Slag (kg/m³)	77	0	0	0
Other Pozzolan (kg/m³)	0	0	50	0
Fine Aggregate (kg/m³)	758	890	818	745
Coarse Aggregate (kg/m³)	1006	890	1011	1030
Water (kg/m³)	166	178	173	165

Performance Characteristics

SCC is a new class of concrete mixture. Its overall performance combines concrete's existing ability to produce a wide range of engineering properties with an increased potential for constructability that exceeds anything possible with conventional concrete. The fresh properties of SCC extend beyond what is traditionally targeted and have expanded the continuum of concrete workability, as presented in Figure 1.1. This extension is quite significant and has the potential to dramatically change concreting practices.

It is not sufficient however, to say that SCC is simply more fluid than conventional slump concrete. SCC will be used differently and placed without vibration; therefore, additional terminology has been developed to describe its fresh performance characteristics more completely. These new characteristics include filling ability, passing ability, and stability. Filling ability refers to the mixture's ability to fill the formwork without vibration, passing ability is the mixture's resistance to aggregate blocking as it flows through obstacles such as tightly spaced reinforcement, and stability refers to the mixture's resistance to segregation and excessive bleeding. These practical characteristics define how SCC will perform in certain applications and are supported by the fundamental rheological characteristics of yield stress and plastic viscosity. The yield stress of a mixture refers to the force required to initiate flow of the mixture, while plastic viscosity refers to the mixture's internal resistance to flow or cohesiveness. Both the practical and rheological characteristics of SCC are discussed more fully in Chapter 3. Similar to current practices that match concrete workability to project requirements and placement techniques, SCC technology must holistically integrate these same project requirements. Chapter 7 provides guidance for selecting SCC performance criteria based on the intended application. It should be noted that some limitations for SCC do exist and are discussed in Chapter 2.

Figure 1.1 Workability Continuum.

The SCC hardened properties of interest are as varied as those for conventional concrete. Extensive investigations have been conducted worldwide on SCC hardened properties, demonstrating that it can be produced with a wide range of compressive strengths, as well as mechanical and durability properties. The fact that a mixture is SCC, and is therefore more fluid than a conventional concrete mixture, is not the primary driver of its hardened properties; however, one SCC property that may influence hardened properties is the mixture stability. Instability can influence the bond to reinforcing steel and durability, among other things. Chapter 4 reviews the details of hardened properties and discusses the variables that drive certain performance characteristics.

History of SCC

Since the late 1990s, the level of investigative activity around SCC has been very high. This has resulted in more than 10 international technical conferences, the publication of over 1000 technical articles on SCC,[12] and numerous graduate student theses. The concept of SCC originated in Japan in the mid 1980s as a result of durability concerns due to the lack of construction labor skilled in the practice of proper consolidation. Professor Hajime Okamura presented the first publication on SCC in 1989 at the Second East-Asia and Pacific Conference on Structural Engineering and Construction.[13] From that time until the present, a number of significant events have occurred, including the following:

- 1997—RILEM Technical Committee (TC 174-SCC) on SCC is formed.
- 1998—International Conference on SCC is held in Kochi, Japan.
- 1999—First International RILEM Symposium on SCC is held in Stockholm, Sweden.
- 2001—ASTM International Subcommittee C 09.47 on Self-Consolidating Concrete is formed.

- 2001—Second International RILEM Symposium on SCC is held in Tokyo, Japan.
- 2002—First North American Conference on SCC is held in Chicago, Illinois, USA.
- 2003—ACI Technical Committee 237—Self-Consolidating Concrete is formed.
- 2003—Third International RILEM Symposium on SCC is held in Reykjavik, Iceland.
- 2005—Fourth International RILEM Symposium and Second North American Conference on SCC are held in Chicago.
- 2005–06—ASTM approves test methods for fluidity, passing ability and segregation resistance.
- 2007—Fifth International RILEM Symposium on SCC is held in Ghent, Belgium.
- 2008—Third North American Conference on SCC is held in Chicago.
- 2010—Sixth International RILEM Symposium and Fourth North American Conference on SCC is held in Montreal, Quebec, Canada.

During this 13-year period, numerous guidelines and test methods have been published globally, including those from Italy, France, Sweden, Norway, and the United States. The following is a partial list of the published consensus or committee-based documents:

- "AIJ Recommended Practice for High Fluidity Concrete for Building Construction", Architectural Institute of Japan, 1997
- "Recommendation for Construction of Self-Compacting Concrete," Japan Society of Civil Engineers, 1998
- "Report rep023: Self-Compacting Concrete—State-of-the-Art Report," RILEM TC 174-SCC, 2000
- "The European Guidelines for Self-Compacting Concrete: Specification, Production and Use," Self-Compacting Concrete European Project Group, 2005
- "Report rep035: Casting of Self-Compacting Concrete—Final Report," RILEM TC 188-CSC, 2006
- "Report rep038: Durability of Self-Compacting Concrete—State-of-the-Art Report," RILEM Technical Committee 205-DSC, 2007
- "ACI 237R-07, Self-Consolidating Concrete," American Concrete Institute, 2007.

Numerous other reports exist besides these.

Acceptance and Development of SCC Technology

SCC is a new innovation for the concrete construction industry, and all innovations go through a process of acceptance and adoption.[14,15] The rate at which an innovation is adopted is influenced by numerous factors, including the social system within which a potential adopter resides. Examples of concrete industry social systems include, but are not limited to, ready-mixed concrete producers, precast concrete producers, contractors, and engineer/specifiers. The innovativeness of the collective group of a social system can impact the innovativeness of the individual within the system. Innovativeness is here defined as how quickly an individual adopts new ideas compared with other members of a system.[1] In this context, both individuals and/or groups can be classified according to five defined adopter categories: innovators, early adopters, early majority, late majority and laggards.

The innovators and the early adopters together make up less than 20% of the overall population of the group being considered. These are the individuals that initiate the change within the group. During the early development of SCC in North America, a small number of practitioners and users formed the initial committees developing standards, testing methods etc. for SCC. Once these initial standards were established and communicated through associations such as RILEM, ASTM, CSA, and ACI, more practitioners and users became involved in these committees, and the acceptance rate increased.

SCC has found a reasonable level of acceptance within the concrete industry. The degree of acceptance is different within the industry groups referred to earlier. For example, approximately 40% of all precast production in the United States occurs with SCC.[16] While in the cast-in-place segment only 2–4% of the concrete placed is SCC.[17] In a recent survey to concrete contractors in North America, 50% of the respondents said that they were aware of SCC, had never used it, but could see its value, indicating that further acceptance in cast-in-place construction is possible.[18] The industry's understanding of SCC has continued to evolve and develop. Tables 1.2 and 1.3 show respectively the number of papers by topic contained in the proceedings of the First International Symposium on SCC and the number of presentations given by topic at the Sixth International Symposium on SCC.

The topics covered since the first conference have become more detailed, based on the previous decade's worth of work. SCC development has evolved from laboratory-focused investigations to include practical research on

Table 1.2 Number of Papers by Topic from the First International Symposium on SCC

Topic	No. of Papers
Rheology/workability	7
Fresh and hardened properties	16
Mix design	8
Constituent materials	16
Applications	18

Table 1.3 Number of Presentations by Topic from the Sixth International Symposium on SCC

Topic	No. of Papers
Rheology/workability	18
Mix design	10
Mechanical properties	9
Physical properties	5
Chemical admixtures	9
Flow modeling	5
Case studies	12
Bond strength	5
Production and placement	10
Fiber-reinforced SCC	10
Formwork pressure	9
Hardened properties	4
Supplementary cementitious materials	5
Durability	5
Aggregates	5
Structural performance	5

different SCC mixture types such as fiber-reinforced SCC as well as investigations of mixing efficiency and other production and placement topics. This all serves to highlight the fact that the state of knowledge around SCC is substantial and has become more refined. This knowledge, although continuing to evolve, is the basis from which continued growth in SCC applications will come. Although research activities will continue and guidelines and test methods will be improved, much of this information now exists, and so the future of SCC is no longer in the hands of researchers, nor is it awaiting the development of test methods or guidelines. SCC's future is now in the hands of the concrete practitioner, waiting for him to put this concrete technology to work.

Summary

SCC is an extension of existing concrete technology and typically uses the same materials as conventional concrete. The primary difference between SCC and conventional concrete is in their fresh property targets. With these new characteristics that are specific to SCC comes the need to understand material interactions, as well as to develop innovative new ways to control SCC's production. Much research has been conducted and the industry understanding of SCC has developed since its inception in the 1980s. Certainly some new developments are needed to further improve its application, but the level of understanding of SCC production and performance are now at the point where the industry can, and currently does, confidently apply this technology on a daily basis. Numerous documents and published papers have provided a significant amount of the information presented in this book, and it is the author's hope that the presentation of this information in combination with some new concepts will provide the concrete practitioner with a solid base from which to implement SCC technology.

Chapter 2

Benefits and Limitations of SCC

Introduction

Courses on marketing and selling typically discuss the difference between the features and benefits of a product or technology. A feature is a description of what a technology does, while a benefit describes how a feature is helpful or advantageous to the user. Not all benefits, however, are of equal importance or value. Additionally, some benefits are easily received with minimal effort, like a gift from a relative, while other benefits require effort to be achieved, such as the improved health benefits of regular exercise. Any benefit, whether given or worked for, requires one to take advantage of it. For example, one could receive a gift from a relative and never open it. The benefit of this gift exists, but is uncaptured. With SCC, the features of high fluidity and self-consolidation result in a variety of potential benefits for the end-user; however, in many cases, effort, such as changing traditional production or placement practices, is required. In addition, the user must plan to take advantage of the benefits. For example, a contractor should plan to redistribute a portion of his or her labor to other activities when using SCC so as to take advantage of its ease of placement benefits. This requires proactive planning on the user's part. This chapter will discuss the potential benefits of SCC and provide a review of benefits received from actual case studies where SCC was used. As with any technology, there are some requirements for and limitations to SCC use that will also be considered.

Benefits of SCC

A clear understanding of the features of a technology is necessary to be able to determine its benefits. It is the author's belief, however, that an earlier presentation of the practical benefits will allow the practitioner to quickly link SCC's features and technical details, presented later, to his or her current challenges and applications. The benefits of SCC have been published or presented on numerous occasions, and the following list is a modified compilation of those benefits from multiple sources:[1-5]

- Reduced human resources needed for placing and consolidating concrete
- More rapid placement of concrete and accelerated construction
- Ease of placement and consolidation in difficult situations owing to access limitations or configuration of element formwork and reinforcement
- Expanded use of concrete in architecturally challenging applications
- Reduction in equipment needs such as vibrators and in some cases concrete pumps
- Reduction in equipment and form maintenance and upkeep
- Shortened concrete delivery times (improved fleet utilization)
- Improved surface finish
- Reduced patching labor and materials
- Improved working conditions for laborers, potentially resulting in:
  o Improved employee retention
  o Reduced employee absence
- Improved safety:
  o Fewer workers on walls needed for placement and consolidation
  o Fewer electrical or air lines running across plant floors for vibration
  o Less noise (especially in precast plants with form vibrators)
- All these have the potential to result in:
  o Fewer injuries and the resulting lost time
  o Reduced workers' compensation claims
  o Reduced insurance premiums.

At some point, the end-user will want to know in hard economic terms what value he or she is receiving by using SCC. It is important to note that the calculation of any benefit will always be relative to what is currently being done and therefore will have to be performed on a case-by-case basis. This process is possible for some of the benefits mentioned earlier such as reduced labor or time for construction. Calculating the financial impact of these benefits is a very straightforward process. However, other benefits may not be as easy to quantify immediately. For example, if a concrete contractor uses SCC on a project due to the presence of complex formwork and the need for an aesthetically pleasing surface finish, it is highly probable that this particular structure is "one of a kind" and therefore the opportunity for the contractor to calculate a relative cost reduction does not exist. In this scenario, the contractor must see SCC as a new tool at his or her disposal to more effectively and satisfactorily complete a project. Therefore, it can be said that SCC is a technology that provides both cost savings and expanded performance benefits.

As previously discussed, a potential benefit must be taken advantage of in order for it to become real. So, in order to learn where SCC is providing real benefits with the greatest frequency, an analysis of published case studies was conducted. Within these case studies, the benefits were described in a variety of ways, and in many cases more than one benefit was highlighted. For example, if the structure was described as having congested reinforcement but the study did not explicitly state that SCC was used for its ability to flow through restricted sections, it was inferred that one of the benefits in this example project was "ease of placement and consolidation through dense reinforcement." Using this logic, the benefits identified in the case studies were assigned to one of the following categories:

- Ease of placement with limited access
- Provide higher in-place quality and aesthetics
- Ease of placement and consolidation through dense reinforcement
- Faster speed of construction and time savings
- Labor savings
- Ease of placement and consolidation in a complex structure or shape
- Improved worker safety and noise reduction.

Table 2.1 provides a complete list of the case studies used in this analysis. The table contains the following information: country in which the project occurred, a brief description of what was cast, whether it was a cast-in-place or precast project, the concrete volume, and the benefits realized based on the above categories. In total, 60 projects are referenced, 50 (83%) are cast-in-place and 10 (17%) are precast.

SCC has been successfully used in many locations around the world and has resulted in a variety of benefits. The majority of published case studies are from cast-in-place projects. This appears to contradict the statements from Chapter 1 regarding the higher level of SCC acceptance within the precast community. However, the greater number of cast-in-place examples is likely a result of the discrete nature of most cast-in-place projects. Precast production is constant, and SCC has been accepted as an everyday tool rather than a project-specific one.

A review of Table 2.1 shows that the benefits of SCC will not be identical for each user or application. In one example, the resource savings may be the greater benefit (this does not mean that quality is not important), while in other cases, such as architectural concrete, the final quality (typically surface finish) is the greatest benefit. For the majority of practitioners, the end-result will be some combination of one or more benefits. The reason for this is that the benefits are based on a "relative improvement." This means an improvement relative to what the user is currently doing, and no two producers or contractors do things in exactly the same way.

It should be noted that the case studies presented and analyzed here are based on those projects whose details have been published and made public. The author is aware of a number of other projects that have never been published and whose details are considered proprietary by the end-user. The reason for highlighting this point is that if all projects that have used SCC

Table 2.1 SCC Case Studies

Ref.	Location	Cast-in-Place (CIP) or Precast	Project	Project Size (m³)	Ease of Placement with Limited Access	Provide Higher In-Place Quality and Aesthetics	Ease of Placement and Consolidation Through Dense Reinforcement	Faster Speed of Construction and Time Savings	Labor Savings	Ease of Placement and Consolidation in a Complex Structure or Shape	Improved Worker Safety and Noise Reduction
6	Japan	CIP	Spillway of dam	4,000	X					X	
6	Japan	CIP	Floor slab of underground structure	4,800	X		X				
6	Japan	CIP	Bottom fill of pneumatic caisson	7,200				X	X		X
6	Japan	CIP	Steel–concrete composite structure	12,000				X	X		
7	Japan	CIP	LNG storage tank	2,230							
8	Sweden	CIP	Strangnas bridges	130		X					X
8	Sweden	CIP	Gavle bridge	200,000		X					X
9	Japan	CIP	Water purification plant			X		X			X
10	Austria	CIP	Columns for Millennium Tower in Vienna		X						
2	USA	CIP	Pedestrian overpass anchor block		X						
2	USA	CIP	National Museum of the American Indian	23,000	X	X	X			X	
2	USA	Precast	Double-tees, Harrisburg Airport parking garage						X		
2	USA	CIP	Rosenthal Center for Contemporary Arts		X	X					
2	Canada	CIP	Reaction Wall, University of Sherbrooke				X				
11	Japan	CIP	Foundation of underground diaphragm walls		X		X				
12	Japan	CIP	MMST tunneling	8,000	X						
12	Japan	CIP	MMST tunneling		X						
13	Korea	CIP	Diaphragm wall for inground LNG tank	32,800	X						

Continued

Table 2.1 Continued

Ref.	Location	Cast-in-Place (CIP) or Precast	Project	Project Size (m^3)	Ease of Placement with Limited Access	Provide Higher In-Place Quality and Aesthetics	Ease of Placement and Consolidation Through Dense Reinforcement	Faster Speed of Construction and Time Savings	Labor Savings	Ease of Placement and Consolidation in a Complex Structure or Shape	Improved Worker Safety and Noise Reduction
14	Canada	CIP	Fill abandoned pump station in mine		X						X
15	Norway	CIP	Rock repository for radioactive waste		X						X
16	Netherlands	CIP	Façade of National Theatre in The Hague			X				X	
16	Netherlands	CIP	Bridge piers		X	X	X			X	
16	Netherlands	CIP	Rotterdam Zoo Hawk Basin walls		X	X	X			X	
17	Canada	CIP	Columns at Pearson International Airport in Toronto	2,100	X						
18	Argentina	CIP	Industrial area walls			X					
18	Argentina	CIP	Bank vault		X		X				
18	Argentina	CIP	Columns in 34-story building		X						
19	USA	Precast	Double-tee production						X		X
20	USA	CIP	Retaining wall	35	X	X	X	X	X		
20	USA	CIP	Footing	92	X		X		X		
20	USA	CIP	Drilled shaft foundation	7	X	X	X				
21	Taiwan	CIP	Building rehabilitation	76	X		X				
22	USA	Precast	Double-tee production					X	X		
23	S Africa	CIP	Nelson Mandela Bridge	86	X	X		X	X		X
23	S Africa	CIP	Bridge 2235		X			X	X		

No.	Country	Type	Application	Quantity							
23	S Africa	CIP	Private residence							X	X
24	Portugal	CIP	Repair of residential building							X	X
24	Portugal	CIP	Repair of wharf structural elements				X			X	X
25	New Zealand	CIP	New winery production facility	96			X				
26	Italy	CIP	St Peter Apostle Church					X	X		
26	Italy	CIP	World Trade Center, San Marino					X	X		
27	Canada	CIP	Retaining wall	100					X		X
28	Chile	CIP	Sewage tunnel	5,500					X		
29	Sweden	CIP	Railway tunnel	2,618				X	X		
30	Italy	CIP	Foundation for cement mill	500					X		X
31	Canada	CIP	Garden City Skyway	979							X
32	Israel	CIP	Intake pumping station walls	900					X		
33	USA	CIP	LNG storage tank	25,000				X	X		X
34	USA	CIP	The Cathedral of Christ the Light in Oakland, CA				X	X	X	X	X
35	Mexico	Precast							X	X	
35	Mexico	Precast							X	X	
35	Mexico	Precast						X	X	X	
35	Mexico	Precast						X	X	X	
35	Mexico	Precast						X	X		
36	USA	CIP	Architectural retaining walls, residential	150–230					X		
37	China	CIP	Railway track slab			X		X		X	X
38	Italy	CIP	Foundations and slabs for housing	123,000		X		X		X	
39	New Zealand	Precast	Precast beams			X		X		X	
40	France	CIP			X			X		X	X

were analyzed for the benefits received, the distribution of the benefits might change. However, it is the author's belief that, at present, this benefits analysis is representative of the current situation and provides some very useful information for the practitioner. Figure 2.1 shows the percentage of time each of the above benefits was achieved out of the 60 case studies listed. The figure also shows the frequency of those benefits by either the cast-in-place or the precast segments.

Figure 2.1 demonstrates how different groups will benefit from SCC. The three primary benefits of SCC for cast-in-place projects are ease of placement with limited access, better in-place quality and aesthetics, and ease of placement and consolidation through dense reinforcement. The three primary benefits of SCC for precast projects are better in-place quality and aesthetics, faster speed of construction, and time and labor savings. The unknown in this analysis is whether the reported benefits are a result of the current acceptance level of SCC within each segment and whether these benefits will change as acceptance grows, in particular within the cast-in-place segment. For example, the speed of construction and the time and labor savings benefits rank very high for precast but quite low for cast-in-place. As will be shown in a later section, significant labor and time savings can be achieved in cast-in-place projects, but this requires concrete contractors to modify their current methods and practices, whereas precast producers have already done this. Therefore, as acceptance of SCC grows in the cast-in-place segment of the industry, these specific benefits will likely be experienced more frequently. In either case, this analysis does show that SCC is particularly useful in difficult placement situations and can provide an economic benefit to the end-user.

Some case studies provided more quantitative information on the received benefits from using SCC. The following sections will review this information in more detail.

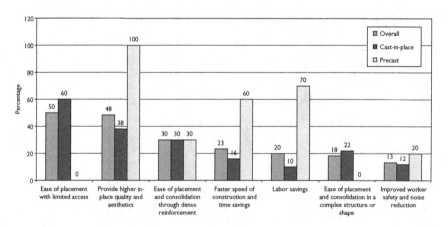

Figure 2.1 Frequency of Specific SCC Benefit by Segment.

Benefits to the Precast Producer

At present, the precast concrete industry has more fully embraced the use of SCC than the cast-in-place segment has. This is for three reasons. First, the number of different mixture proportions produced daily is limited, increasing repetition, and therefore making quality control relatively easier. Second, the conventional mixtures used in precast/prestressed production are typically proportioned to achieve higher early compressive strength development, and therefore the relative material cost increase to produce an SCC mixture is not as high compared with other scenarios. Finally, in most cases, the precast producer has the advantage of being both the concrete producer and the contractor/placer. In this scenario, most of the benefits of SCC are captured internally and any cost increase for SCC is due only to the increase in material costs, making it easier to justify.

The data provided in Table 2.2 come from two previously published studies of the value of SCC in precast production.[41,42] Both studies used the production of prestressed double-tees as their basis. The data are presented as a comparison of labor and resources for casting with conventional concrete versus SCC. Study 1 was based on a full day's production and categorized labor by the following activities: batching, transport, placing, vibration, floating/finishing, clean-up, and patching.[41] Study 2 compared the pouring time/labor and patching time/labor for a single bed's production.[42] In order

Table 2.2 Placement and Patching Labor Comparison of SCC with Conventional Concrete

	Study 1			Study 2		
	Conventional	SCC	% Reduction	Conventional	SCC	% Reduction
Placement/ Consolidation Labor (people)	13	11	15	14	7.5	46
Placement/ Consolidation Time (hours)	2.5	2	20	11	9	18
Placement/ Consolidation Total Man-Hours	32.5	22	32	154	67.5	56
Patching Labor (people)	2	1	50	6	1	83
Patching Time (hours)	8	9.5	−19	11	9	18
Patching Total Man-Hours	16	9.5	41	66	9	86

to make these two studies more directly comparable, only the placement/ consolidation and patching items were used. Patching refers to the process by which the as-cast surface of an element is made acceptable by applying a mortar mixture to fill in small surface blemishes entrapped during casting, Figure 2.2 shows the surface of a partially patched element. The left side has not been patched, while the right side has.

Table 2.2 shows the reduction in both labor and time for both studies. To establish a rough rule of thumb, we can use the average of these two studies and say that placement and consolidation labor costs will be reduced by 40% and patching labor costs by 60%. These numbers can provide a starting point for estimating the value of SCC used in precast production.

The reference document for Study 2[42] provided daily labor cost savings of $3933.00/day and the labor compensation rate was $23.00/h. When casting an average of 495 m^3 (650 yd^3) of concrete per day, this is equivalent to $7.95/m^3 ($6.05/yd^3) in labor cost savings via the use of SCC.

Another source of data reported that for two precast producers, one producing utility structures and the other producing water-retaining structures, the in-place savings were $9.73/m^3 ($7.40/yd^3) and $12.65/m^3 ($9.61/yd^3), respectively. These values included the increase in material cost and the decrease in equipment and maintenance and labor. Annual labor savings were

Figure 2.2 Partially Patched Formed Surface Finish.

$100,000 and $140,000 and the reduction in equipment and maintenance was $15,000 and $20,000, respectively.[43]

There are also material cost savings for patching. If we assume that each laborer assigned to patch surface imperfections works at the same rate in terms of area patched per hour, we can then use the labor hour reduction percentage as an estimate of patching material savings. For example, Study 1 above would estimate a 41% savings in patching material while Study 2 would estimate an 86% reduction in the amount of patching material used. This can be substantial, depending on the type and cost of the material used for patching.

Other benefits to precast concrete producers (and concrete contractors) include the health and safety of workers. These benefits can come from four primary areas: noise reduction, absence of vibration, ergonomics and physical strain, and hazard reduction. For example, form and table vibrators, which are often used in precast production, can be quite loud. One study showed that for an individual within 1 m of these vibrators when running, the noise level was over 102 dB. When internal vibrators were used (which is most often the case for cast-in-place projects), the noise level was approximately 82 dB, and when no vibrators were running, the background noise in these studies ranged from approximately 60 to 80 dB.[44] The use of SCC could then significantly reduce the noise level in a precast plant or on a construction site.

With extended use of hand vibrators, certain ailments can befall workers, including hand–arm vibration syndrome resulting in vibration "white fingers." This is a condition considered similar to Raynaud's syndrome. The physical strain of pulling, raking, and vibrating conventional concrete is also quite high. Consider a worker standing on top of a form lowering and raising a poker vibrator into and out of the concrete time after time for several hours (Figure 2.3) or a concrete laborer raking concrete into place. The use of SCC completely eliminates this activity. From these benefits the precast producer may experience value in having a more content workforce resulting in fewer called in sick days.

Figure 2.3 also illustrates the safety issues associated with consolidating concrete. In this photograph, the worker is standing on a form that has been coated with a form release, numerous clamps are sticking up from the formwork, creating a tripping hazard, the vibrator cord is creating another tripping hazard, and there is no fall protection. The use of SCC would eliminate all of these safety hazards associated with concrete consolidation and would therefore potentially reduce the instances of worker injury, lost-time accidents, and workers' compensation claims. The monetary impact of these workplace benefits is difficult to quantify, which is one reason why they are not reported as often as other benefits in the case studies referenced. It is the author's opinion, however, that as many corporations in North America continue to devote significant time and resources to workplace safety, these benefits will become more important and more valued in the future.

Figure 2.3 Ergonomics and Safety of Concrete Consolidation.

A reduction in equipment maintenance costs and a reduction or delay in capital expenses due to stress on the mixing and handling equipment are other potential benefits of SCC to the precast producer. For example, because of SCC's fluidity, the power draw on the mixer motor and the stress on the equipment can be less with SCC than with a conventional slump mixture. Forms should last longer through the use of SCC, in particular in those factories that employ surface vibrators attached directly to the forms. Eliminating vibration will eliminate this particular stress on the forms and reduce the frequency of repair, while potentially delaying the forms' replacement. These form surface vibrators are usually run by compressed air, and the airlines are run along the walls and across the floors of the precast factories. Elimination of these airlines improves safety and removes an item needing maintenance, including the air compressor itself. These compressors can have significant loads applied to them during daily production, and if one breaks down, production interruptions can occur. The energy costs associated with the creation and maintenance of compressed air used for consolidation can be reduced or eliminated by the use of SCC. What this amounts to in monetary

Table 2.3 Percent Savings Due to the Elimination of Vibration

	Savings
Energy Consumption	10%
Form Costs	20%
Maintenance Costs	10%
Illness Time	10%

terms is dependent upon the specific precast factory; however, Table 2.3 shows the potential savings due to elimination of vibration.[45]

Benefits Specific to Ready-Mixed Concrete Producers[46]

A ready-mixed concrete producer can benefit from SCC in the following ways: improved fleet utilization, less wear and tear on equipment, and creation of a new, value-added, offering for customers.

Can SCC reduce costs for the ready-mixed concrete producer? In the area of fleet utilization, there may be an opportunity. In conversations with a number of concrete producers, the topic of fleet utilization and truck time has been communicated as a key performance indicator (KPI) for many companies. In the course of these conversations, it has been revealed that the time cost of a fully loaded ready-mixed concrete truck can range from $0.50 to $2.00 per minute. SCC can impact this time cost for a truck in two ways: first by the rate at which concrete can be discharged from the truck itself, and second because SCC allows a contractor to place concrete more rapidly, thereby unloading the truck more quickly. The size of this benefit will be variable, depending upon the project. For example, in the casting of a typical residential slab with SCC, the rate at which the truck can be unloaded is not limited by the contractor, assuming a fairly open slab area. The truck can simply pull up to the slab and unload. However, in a case such as the casting of columns using a crane and bucket, the rate of discharge is governed by the contractor. In this latter case, the contractor should see time savings due to faster discharge from the bucket and the removal of the requirement for vibration. In both cases, the time to unload the truck should be reduced; however, the amount of time saved will be different. To make these savings real, however, the concrete producer must evaluate and potentially change the logistics plan for servicing SCC jobs.

One estimate reported that the use of SCC could increase the concrete placement rate from 25–35 m^3 per hour to 50–315 m^3 per hour, depending on the project and the placement techniques used. For the concrete producer, this allows the unloading of 5.5–35 trucks an hour with SCC compared with 3–3.5 trucks per hour with conventional concrete. This can produce a completely different logistical scenario compared with what is currently done.[47]

Benefits to the Concrete Contractor

The general list of potential benefits at the beginning of this chapter reveals that many of them apply to the end-user or the party responsible for placement of the concrete. For cast-in-place projects, this would be the concrete contractor. Worker safety and health issues from the contractor's perspective were covered above in the section on precast concrete and are not repeated here. Let us then consider the benefit of easier and faster placement potential. What does this mean? It means that the contractor should be in a position to do two things: reduce the number of laborers dedicated to placement and consolidation activities and complete the project faster. Table 2.4 provides three real-world examples where SCC was compared with conventional concrete in terms of placement time and labor required.[48] Data are provided on the project, placement technique, and project concrete volume, and on the percent reduction in placement time, placement laborers, and overall placement man-hours when SCC was used versus conventional concrete.

In a second study, SCC was compared with the projected/theoretical requirements of conventional concrete. The data presented represent two bridge construction projects and two separate pours for each project. Table 2.5 shows the project, project concrete volume, and percent reduction in placement time, placement laborers, and overall placement man-hours when SCC was used versus conventional concrete.[49]

Table 2.4 Savings in Placement Time and Labor with the Use of SCC for Three Projects

Project	Placement Technique	Volume (m³)	Percent Reduction		
			Placement Time (hours)	Placement Labor (people)	Placement Man-Hours
Retaining wall	Pump	35	36	50	68
Footing	Pump	92	50	60	80
Drilled shaft foundation	Chute	7	66	50	83

Table 2.5 Savings in Placement Time and Labor with the Use of SCC for Two Projects and Two Pours for Each Project

Project	Volume (m³)	Percent Reduction		
		Placement Time (hours)	Placement Labor (people)	Placement Man-Hours
Bridge 1 superstructure	47	0	75	75
Bridge 1 superstructure	90	33	75	83
Bridge 2 superstructure	207	30	0	30
Bridge 2 superstructure	207	37	0	37

In reviewing Tables 2.4 and 2.5, a reduction in total man-hours of 30% to over 80% seems possible and placement times can be reduced by up to 66%. The placement time saving can provide further savings on rental of equipment such as concrete pumps. It can also benefit the contractor and/or owner by resulting in faster project completion (depending on the size of the concrete placement portion of the project). At the very least, it can provide a time buffer for the contractor with respect to the expected completion date.

An improved as-cast surface appearance is frequently referenced (48% of the time in the case studies analyzed) as a benefit to using SCC. This improvement to the surface finish results in reductions in patching labor and materials costs. Although no studies quantifying this cost savings benefit in cast-in-place applications have been published to the author's knowledge, several case studies have provided a qualitative description of the savings (see Table 2.1), indicating that these costs are either eliminated or significantly reduced. As a starting point, it would be reasonable to adopt the patching labor savings figures provided in Table 2.2 for precast concrete, where the costs have been shown to drop by 40% to more than 80%. One source reports that the cost of patching a vertical surface with cement mortar (not prepackaged mortar) is approximately $11.75/m² ($1.10/ft²).[50]

Another published study on the benefits of SCC relative to conventional concrete in cast-in-place construction estimates a 6–10% saving on the overall concrete operation costs when using SCC, and, similar to precast concrete production, the overall improvements to the working condition of concrete laborers can be significant. The noise levels that laborers are exposed to are approximately one-tenth of those produced when vibrating conventional concrete. With respect to ergonomics, this study reports that the effort required to place traditional vibrated concrete was rated as hard, while that for placing SCC was rated as fairly light.[51]

Specific benefits also exist in flatwork construction. Specific advantages surround the absence of the need to provide a final finish on the slab. This results in a labor reduction of almost 50% and reduces the placement time significantly. A reduction in placement time provides an advantage when the project is constrained by local noise bylaws that limit the duration over which construction activities can occur. This is especially true in areas with a dense residential or institutional population. This is limited to floors needing a floor flatness of 20–25. If greater floor flatness is needed in these situations, then a thin scratch coat may be required.[52]

Finally, a case study from Japan for the construction of an LNG storage tank reported that the number of concrete laborers on the project decreased from 150 to 50, and the time required for construction of the structure decreased from 22 months to 18 months. Additionally, on another large project in Japan, the construction of the huge anchorage for the Akashi–Kaikyo bridge, construction time was reduced from 30 to 24 months.[53]

Based on these studies, there appears to be significant opportunity for cost savings when using SCC. In order to make the potential benefits real, however, the user will need to do something different, such as schedule fewer laborers for an SCC project compared with a conventional concrete project. The concrete contractor then needs to estimate how much time saving can really be gained by using SCC and scheduling the project activities differently. In the cases presented, typically both the time to cast and the amount of labor required were reduced, which means that potentially rather than not having the labor on the jobsite, those who would originally have been designated for concrete placement can now be redirected to other activities to further accelerate the project completion. Based on the above, SCC has the potential to dramatically change both the process and the end-result of cast-in-place concrete construction.

Limitations of SCC

As with any technology, there are some constraints on its application. These can be categorized in two ways: project and organizational constraints. Project constraints refer to those of a technical or economic nature. Organizational constraints focus specifically on the abilities of the concrete producer, contractor, and/or specifier to effectively apply SCC technology to their portion of the concrete construction process.

Project Constraints

SCC is meant to be highly fluid; therefore, projects where the concrete is required to immediately support its own weight or to hold a significant grade may not benefit from its use. The following specific examples may help to illustrate these limitations more clearly.

Example 1

Areas where the concrete is required to hold some amount of slope are one example of applications imposing limits on the use of SCC. Billberg et al.[54] reference three bridge projects where SCC was used for the entire structure, including the deck. On two of the bridges, the deck required a 2.5% slope, while on the other a 4% slope was required. The original mixtures for these projects targeted 650 mm (26 in.) slump flow. Although the reference does not say what the slump flow was for the mixtures used on the concrete deck, it does say that it was lower, presumably to allow the concrete to hold the required slopes. The slump flow will be limited to a certain maximum value for projects requiring a slope to be held. It should also be noted that an SCC mixture being used on a sloped surface should be tested for its response to vibration, particularly those levels of vibration common to a construction site. As vibration tends to cause mixtures to behave as if their yield stress has decreased, these mixtures could sag as vibration occurs.

Example 2

During the construction of many sports stadiums, precast seat risers are used. In most instances, these risers are cast in forms, with the vertical portion completely formed and the horizontal portion formed on the bottom and sides but with the top open (Figure 2.4). If a highly fluid concrete mixture is placed into the vertically formed section, the concrete will flow out of the open horizontal section. One way that this can be overcome is to proportion a mixture with a controlled workability retention time so that it stiffens rapidly enough to match the casting process. If this balance is not achieved, then a highly fluid SCC mixture will not stay in the open horizontal section, and SCC is therefore not an appropriate solution. Another way to overcome this limitation is to completely close the open faces and pour the entire piece.

Organizational Constraints

Organizational constraints include those of the concrete producer, contractor, and/or the specifier. Some concrete producers may be limited in their ability to consistently produce good-quality SCC. These limitations are not due to a lack of intent or desire, but rather to the time, equipment, and resources needed to fully implement SCC or any other high-performance concrete technology. For example, is the production equipment sufficient for moisture

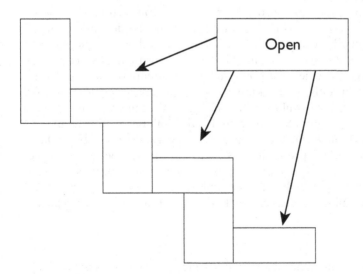

Figure 2.4 Basic Schematic of a Stadium Riser.

and fluidity control? In addition, areas such as quality control, mixture development, and production logistics may need modification or expanded resources when SCC is used. Is the producer willing to invest resources in order to make these modifications? A producer must be committed to SCC for there to be lasting success. The commitment will come in terms of the upfront time spent on equipment review, material investigation, mixture development, personnel training, and quality control. Many of these items are further discussed in Chapters 7–14.

Delivery logistics of SCC requires special attention. Consider the increased rate at which SCC can be placed relative to conventional concrete. In Tables 2.4 and 2.5 the placement rates are shown to be reduced by 30–70% when using SCC versus conventional concrete. This places an additional burden on the concrete producer regarding the staging of trucks, particularly at the beginning of the pour. Another delivery limitation comes from the highly fluid nature of the SCC mixture. Most concrete mixer trucks do not have lids that close the discharge opening. This means that there is potential for a fluid mixture to spill out of the back of the truck if the truck is batched to its maximum capacity, particularly in hilly terrain. In these cases, trucks may be batched to a lower percentage of their capacity. This will affect the logistics time as well as the fleet utilization benefits mentioned earlier. These things should be considered by the concrete producer during the project planning stages. One solution to this last limitation is to batch the mixture to a lower consistency and add the remaining high-range water reducer (HRWR) on site, bringing it to the appropriate SCC consistency. This, however, would require the positioning of a dedicated technical resource on site to handle the admixture dosing activities.

Organizational limitations also apply to concrete contractors. SCC should be viewed as a new tool to be used appropriately. To do this, proactive planning is required, particularly for the first-time user. For example, consider a project where a slump of 200 mm (8 in.) is typically used, and then assume that the contractor decides to use a 50 mm (2 in.) slump for no reason other than preference. Intuitively, one knows that the process for deploying labor, equipment, timing, etc. would need to change. In the same way, if one were to substitute a highly fluid SCC mixture for the 200 mm (8 in.) slump mixture, the methods and techniques for placing and finishing should be modified to take advantage of the properties specific to SCC. If one goes into a project using SCC and has not planned on how to appropriately take advantage (technically and economically) of this mixture, then the resulting benefits will be significantly limited. In this case, lack of preparation is the limiting factor.

Summary

This chapter has shown that the available benefits of SCC are similar across projects regardless of whether one is producing precast segments or casting architectural walls. However, the specific benefits actually obtained will be just as individual as the projects themselves and the producer or contractor involved. The financial impact of the benefits will always be relative to what is currently done and within the context of where the work is taking place. This is especially true when calculating the actual financial impact of the time and labor benefit. This must be done locally and within a segment due to the wage scale differences. For example, one source reported labor cost as $22.00/h for precast labor, while another reported it as $65.00/h for cast-in-place labor.[43,47] This difference can make a significant impact on the value of SCC. In this sense, it will always be up to the user to determine his or her own value proposition. Additionally, the benefits may not come from only a cost savings perspective. In some cases, the real benefit may be a function of the end-result, such as greater in-place quality and surface finish that could not have been achieved otherwise. But, as with any potential benefit, it is up to the individual to make it real. The constraints that were highlighted are not overly burdensome or difficult to handle; however, they should still be consciously considered by the producer and contractor before initiating an SCC program. Overcoming these initial constraints will enhance the potential for future success in implementing SCC technology.

Chapter 3

Fresh Properties and Test Methods

Introduction

The fresh concrete properties of importance for any discussion of SCC are those that influence its placement and consolidation. It is in these properties that SCC shines and differentiates itself from other concrete mixtures. This is not to say that its hardened properties can or should be ignored. They should not be, as they are just as important with SCC as for conventional concrete. Most current concrete technologists, however, have a reasonable understanding of how to adjust proportions to achieve certain hardened properties, and if they do not, there are well-established guidelines or other resources that can readily explain these methodologies. That level of industry-wide understanding does not yet exist with SCC and its fresh properties because new performance characteristics that many have not heard of or experienced have been created in order to define SCC performance. It is important therefore to have a clear understanding of the fresh properties that make SCC different, to know why they are important, how they interact with one another, and how they are measured. The key to SCC is control and balance of its fresh properties.

This chapter will describe the fresh properties that are fundamental to SCC and the constituent materials that influence them. A brief discussion of the importance of the science of rheology will also be given. The common methods used to measure these properties will also be reviewed.

SCC and Concrete Rheology

SCC mixtures are much more fluid than conventional concrete mixtures. Therefore, the manner in which they are characterized requires a different perspective and new measurement techniques. Rheology is the science dealing with the flow of materials.[1] Although the fresh properties of concrete were described in rheological terms by Tattersall and Banfill[2] in the early 1980s, SCC has now provided the industry with a concrete whose very foundation was built on rheological characteristics. With SCC, the flowing properties of a concrete mixture are paramount and rheology helps us to both understand

and differentiate performance in a fundamental way. Numerous rheological models exist, but most documents suggest that concrete should be rheologically described according to the Bingham model.[3,4] This model proposes two constants defining the flow of a material: the yield stress, which is defined as the amount of force required to initiate flow of a material, and the plastic viscosity, defined as the material's internal resistance to flow. For concrete, these parameters are measured through the use of a concrete rheometer, several of which are commercially available. Although these rheometers will rank the relative performance of compared mixtures similarly, the values generated by each will be different,[5,6] and this makes it difficult to compare data generated in different laboratories using different rheometers. However, they are still very useful for studying concrete performance. The typical concrete rheometer will measure torque values as a sample of concrete is sheared at different rates. This type of measurement sequence will provide a plot similar to that shown in Figure 3.1.

The plot in Figure 3.1 shows the increased torque on the impeller as the rate at which it spins through the concrete increases. This would be analogous to the stress on one's arm while mixing cake batter faster and faster in a mixing bowl. The faster one mixes, the more stress is applied to one's arm and the mixing becomes more difficult. The rheological data is derived by plotting a best-fit line through this set of points. The extrapolated point at which this line intersects the y-axis is the yield stress of the material, which is the point at which the mixture will "yield" or begin to flow. Applied stress levels above this point will cause the concrete to flow, but below this level the concrete will

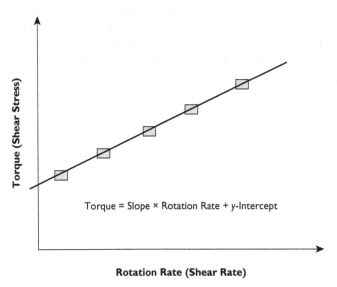

Figure 3.1 Example Output from a Concrete Rheometer.

not move. The yield stress is inversely proportional to the slump of the concrete. A comparison of slump versus yield stress can be seen in Figure 3.2. The rheological data in this figure was produced using the IBB Rheometer.[6]

Consider two conventional concrete mixtures: one low-slump and one high-slump. A higher stress (more effort) must be applied to a low-slump concrete mixture to move it during placement, compared with a high-slump mixture, where the stress required to initiate movement is lower. For SCC, the yield stress should be very low so that the material flows exclusively under its own weight and the techniques required for placing and consolidating conventional concrete are not required. For this reason, traditional measurement techniques such as slump are not adequate to characterize SCC. When measuring strictly the slump of SCC, the value will always be very high, with no ability to discriminate between mixtures.

As stated previously, in addition to the yield stress, the plastic viscosity of a concrete mixture can be quantified using a concrete rheometer. ACI 238 defines viscosity as ". . . the resistance of a fluid to deform under shear stress." What this means to the concrete practitioner is that the mixture, although more or less "fluid," becomes more or less sticky. Consider the difference between water and honey: both are fluids with zero, or close to zero, yield stress, yet honey has a higher viscosity than water. The higher viscosity of honey-like fluids will result in two things important to SCC: they will flow more slowly than water-like fluids, although the ultimate flowing distance may be similar, and movement of solid materials through honey-like fluids will be more difficult than in water-like fluids. The viscosity of SCC mixtures can vary significantly depending upon materials, mixture proportions, and admixtures. Some mixtures can have higher plastic viscosity, like honey, while others are lower. This is important as it can impact certain performance attributes such as segregation resistance.[7] Rheologically then, SCC mixtures are characterized as having a low yield stress and a plastic viscosity that varies

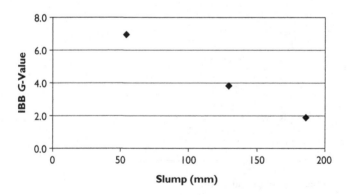

Figure 3.2 IBB G-Value versus Concrete Slump.

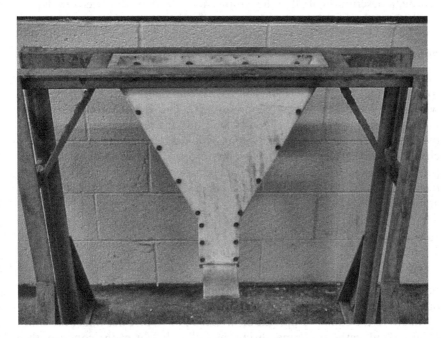

Figure 3.3 V-Funnel Apparatus.

with the intended application.[7] Aside from using a concrete rheometer, other test methods exist that can provide the practitioner with a relative indication of the plastic viscosity, one of which is the V-funnel test described below.

The V-funnel test[8] is performed by filling a V-shaped funnel (Figure 3.3) with SCC, opening the trap door on the bottom, and recording the time it takes for all of the SCC to flow out of the funnel. This test has been correlated to concrete viscosity, and a longer V-funnel time, as long as no blocking occurs, indicates a higher plastic viscosity.[6,9]

Another test for viscosity, the T50 test, will be described, along with the slump flow test for filling ability, in the following section on SCC fresh properties.

SCC Fresh Properties

Most concrete practitioners do not have access to a concrete rheometer, and therefore other characteristics that are both more easily measurable and useful for describing practical SCC performance have been established. The three primary characteristics that define SCC performance are filling ability, passing ability, and stability. This section will define each characteristic and then present the test methods typically used to measure them. The methods presented will be those that are standardized and regularly used. A more thorough description of all of the various methods that have been developed for SCC can be found in other documents.[8,10–12]

Filling Ability

Filling ability is the ability of the fresh concrete mixture to flow into and fill formwork under its own weight. This term is sometimes used interchangeably with fluidity and flow as they pertain to SCC. This is the characteristic that most will recognize when they see an SCC mixture flowing freely, under its own weight, across a form. Most people would agree that this is the primary characteristic that defines SCC. The other properties to be mentioned can be present to a greater or lesser degree depending upon the application requirements, but for a mixture to be considered SCC, it must have adequate filling ability.

Test Methods

SLUMP FLOW TEST[13]

In performing this test, the standard Abrams cone is filled in a single lift without rodding, because the concrete consolidates on its own. The cone can be set in either the standard upright or inverted position. Once filled, the cone is raised and, rather than measuring the falling height of the concrete, the diameter of the resulting concrete paddy is measured. See Figure 3.4.

In addition to the spread, other measurements and observations are taken. From the instant the cone is lifted, the time the spread takes to reach 50 cm (20 in.) is measured. In the literature, this measurement is referred to as T20,

Figure 3.4 Slump Flow Test. (Courtesy BASF Construction Chemicals, Photographer Hal Stata.)

T50 or T500 for inches, centimeters, and millimeters respectively. T50 is the most common term used, however. For mixtures with similar final slump flow values, this flow time can provide a relative indication of the mixture's viscosity. A higher T50 time in general indicates a higher viscosity as long as final slump flow values are similar. Another very important parameter that can be assessed along with the slump flow is the visual stability index (VSI) rating. The VSI is further described in the section on stability below.

Passing Ability

This refers to the ability of an SCC mixture to flow through restricted spaces without blocking. This property is generally concerned with aggregate flowing through reinforcement; however, it can also refer to flow through narrowing sections in formwork or when reducers are present on concrete pump lines. Any situation where the solid particles in the concrete have to rearrange themselves in order to flow through an obstacle is where passing ability characteristics are important (Figure 3.5).

Test Methods

J-RING TEST[14]

This test consists of performing a slump flow test with a ring of reinforcing bar around the base of the cone (Figure 3.6). There are various ways to

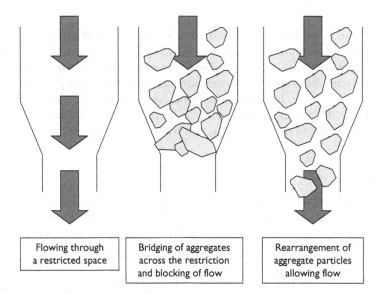

| Flowing through a restricted space | Bridging of aggregates across the restriction and blocking of flow | Rearrangement of aggregate particles allowing flow |

Figure 3.5 Examples of Aggregate Blocking and Flowing through a Restricted Space.

Figure 3.6 J-Ring Test.

quantify the value from this test, either by measuring the height difference inside and outside of the ring,[15] or, as in the ASTM method, by calculating the difference between the slump flow value with and without the ring. The smaller the difference between the two measurements the better the passing ability of the SCC mixture. In addition, a visual assessment of the concrete inside and outside of the ring can also provide an indication of stability or resistance to separation of the mortar fraction from the coarse aggregate.

L-BOX TEST[8,15]

This test consists of an L-shaped box with a sliding gate separating the vertical and horizontal sections of the box (Figure 3.7). On the outside of the gate is a set of reinforcing bars through which the concrete flows as it passes from the vertical to the horizontal section of the box. The ratio of the height at the opposite end of the horizontal section to the height in the vertical section is calculated. The larger the number, the better the passing ability. The speed of flow through the horizontal section can also be measured, giving an indication of mixture viscosity.

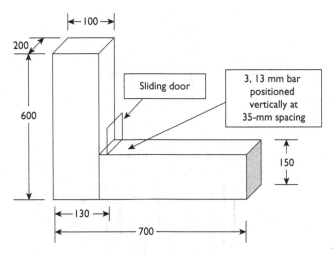

Figure 3.7 L-Box Test (all dimensions in mm).

U-BOX TEST[8,16]

This test consists of a U-shaped box separated down the middle by a thin wall and a sliding gate. At the bottom of the wall, one of two different configurations of rebar can be placed as obstacles: one with more rebar and narrower spacing (Rank 1) and the other with slightly more open spacing (Rank 2) (Figure 3.8). Concrete is placed on one side of the box, the sliding gate is opened, and the concrete is required to flow from one section to the other through the rebar obstacle. The rising height on the once-empty side is measured: the greater the height, the better the passing ability. This test can also be run without the rebar obstacle present (Rank 3). The Japan Society of Civil Engineers recommends a minimum rising height of 300 mm (11.8 in.).

Stability/Segregation Resistance

This characteristic refers to the ability of an SCC mixture to resist separation of its constituent materials. Stability is determined under two conditions: dynamic and static. Dynamic stability refers to the segregation resistance of a mixture during transport, placement, and up to the point where static stability takes over. Two examples of this are stability during transport in non-agitating equipment and stability during long flowing distances. In many precast factories, the concrete is delivered to the forms in vessels without agitating equipment. If the mixture is not designed appropriately, settlement of aggregate particles can occur due to the jostling and vibrations during transport. In addition, if a mixture is not proportioned correctly and the

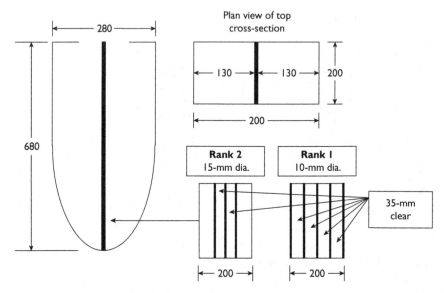

Figure 3.8 U-Box Test and Obstacles (all dimensions in mm).

concrete is allowed to flow for too long a distance, the fine mortar fraction has been seen to separate from the bulk of the concrete. Techniques for improving the dynamic stability of SCC include reducing the slump flow level and reducing the coarse aggregate size and density.[17] As yet no standardized test methods have been developed that measure dynamic stability.

Static stability refers to the ability of the mixture constituents to resist segregation and settlement as the concrete sits undisturbed. This could take the form of coarse aggregate particles settling or water and air migration towards the surface.

Test Methods

COLUMN SEGREGATION TEST[8,18]

In this test, a 650 mm (26 in.) column is used. It is constructed so that the top and bottom quarters can be removed. During the procedure, the column is filled with fresh SCC and allowed to stand undisturbed for 15 minutes (Figure 3.9). At the end of this time, the concrete from the top and bottom quarter sections is removed and washed over a 4.75 mm (#4) sieve so that only coarse aggregate remains. The difference in coarse aggregate content between the top and the bottom is determined and the percent segregation is calculated.

Figure 3.9 Column Segregation Test.

SIEVE SEGREGATION RESISTANCE TEST[8,16]

This test consists of collecting a sample of fresh SCC, allowing it to sit in a bucket undisturbed for 15 minutes, then pouring the top section of the sample onto a sieve, thereby allowing any water, paste, and mortar to flow into a receptacle. The amount of material that passes through the sieve is calculated as a percent of the original sample obtained.

PENETRATION APPARATUS TEST[8,19]

In this test, an empty tube is set on top of a freshly cast sample of SCC and the penetration depth of this tube after a given time is recorded. In general, the tube penetrates until it contacts coarse aggregate; as coarse aggregate continues to settle, the tube will penetrate further. Figure 3.10 shows a picture of the penetration apparatus. This test has been calibrated to the column segregation test and is much more useful for field testing of stability.

VISUAL STABILITY INDEX (VSI)[8,13]

This test is conducted as part of the slump flow test mentioned previously. It consists of making a visual observation of the slump flow paddy and assigning to that observation a number between 0 and 3, designating the

Figure 3.10 Penetration Apparatus for Stability Testing.

degree of instability. A mixture with a VSI rating of 0 would be highly stable, while a mixture with a rating of 3 would be highly unstable. Figures 3.11–3.14 provide examples of the four VSI levels. The original VSI process[20,21] also recommended to observe the concrete in the wheelbarrow and the mixer (if possible). This is important, as there is greater depth of concrete in the wheelbarrow than in the slump flow paddy and settlement may be more readily observable here. The test was originally intended as a secondary means of jobsite control for SCC stability rather than as a method for quantifying performance in a laboratory. Many times, one can see clear evidence of segregation in the wheelbarrow or in the slump flow paddy, and the test was meant to provide a way of categorizing what was seen. Typically a VSI rating of 0 or 1 is acceptable. In general, mixtures with VSI ratings of 0 or 3, which are the extremes, tend to be the easiest to identify. Ranking a mixture as having a VSI rating of 1 or 2 requires more experience with SCC. This is why, in some cases, pictures are posted in the quality control laboratory to assist technicians in assigning VSI values to mixtures.

Figure 3.11 Visual Stability Index (VSI) Rating 0. (Courtesy BASF Construction Chemicals, Photographer Hal Stata.)

Figure 3.12 VSI Rating 1. (Courtesy BASF Construction Chemicals, Photographer Hal Stata.)

Figure 3.13 VSI Rating 2. (Courtesy BASF Construction Chemicals, Photographer Hal Stata.)

Figure 3.14 VSI Rating 3. (Courtesy BASF Construction Chemicals, Photographer Hal Stata.)

BLEEDING[22]

The measurement of bleeding in conventional concrete is a fairly standard test and can be used for SCC as well. Bleed is another facet of SCC stability. Some SCC mixtures will bleed more than others, based on the materials, admixtures, and mixture proportions used. However, the absence of bleeding does not guarantee that a mixture is stable, because coarse aggregate settlement can still occur regardless of whether or not bleeding is present.

Factors Influencing the Fresh Properties of SCC

Numerous material and mixture proportioning factors impact fresh SCC properties. Table 3.1 outlines a general relationship between these factors and the more practical properties of SCC. ACI 238.1R provides a detailed description of how materials influence the rheological characteristics of concrete. It should be noted that this table provides very general relationships, and there is a more detailed discussion of proportioning and material constituents and their effects in Chapters 5 and 8.

Table 3.1 Material Characteristics and Proportioning Influence on Fresh SCC Properties

	Constituent Material Characteristics and Effects		
	Filling Ability	Passing Ability	Stability
Powders	Packing density, particle shape (angular vs rounded) and reactivity of the powder combinations used can impact the water and/ or admixture demand of an SCC mixture.	Minimal to no impact on the passing ability of an SCC mixture: only in so much as they would affect filling ability and stability.	Fineness will impact bleeding characteristics and viscosity. Viscosity level will affect aggregate settlements.
Aggregates	Optimized packing density and more rounded particles will increase the filling ability of an SCC mixture.	More angular particle shape will reduce the ability of particles to rearrange their relative position while flowing through a restricted space. Smaller particles will flow through restricted spaces more easily.	Smaller particles will have lower tendency to settle and segregate. Higher fine contents will reduce bleeding.
High-Range Water Reducer (HRWR)	Increases the fluidity of the paste fraction of the concrete mixture. In a properly designed SCC mixture, this higher-fluidity paste will enhance the properties of filling ability, passing ability, and stability. In a poorly designed SCC mixture, such as one with inadequate paste volume, an overdose of HRWR can occur and result in reduced stability, increased bleeding and poor passing ability.		

Continued

Table 3.1 Continued

Constituent Material Characteristics and Effects		
Filling Ability	Passing Ability	Stability

	Filling Ability	Passing Ability	Stability
Viscosity-Modifying Admixture (VMA)	Increases the viscosity of the paste fraction of an SCC mixture. In a properly proportioned SCC mixture, this will have minimal influence on the filling ability, while enhancing the passing ability and stability. In a poorly designed mixture, this increased paste viscosity can lead to a reduction in filling and passing ability.		

Proportioning Effects		
Filling Ability	Passing Ability	Stability

	Filling Ability	Passing Ability	Stability
Powders	Within a reasonable water/powder ratio range, an increase in powder content will increase the paste volume and paste viscosity, enhancing the filling ability, passing ability, and stability of an SCC mixture. Too great an increase in powder without an increase in water can lead to mixtures with viscosities that are too high, resulting in reduced filling and passing abilities.		
Aggregates	Too high an aggregate volume relative to the paste volume will decrease filling ability.	Too high a concentration of larger particle sizes will cause blocking and reduce passing ability. The concentration is governed by the particle size distribution (gradation) and mixture proportions together.	Higher fine aggregate contents will reduce bleeding and aggregate settlement.
Water	Increased water increases the paste volume and the filling ability as long as no segregation is occurring.	In some instances, an increase in water can increase the passing ability by increasing the paste volume and improving the filling ability. In other cases, it can reduce the paste viscosity too much and decrease the passing ability.	Too high a level of water can increase bleed and instability.

Interrelationship between SCC Characteristics and Testing Methods

Even when making just a single change to an SCC mixture, it is virtually impossible to change only one performance parameter. Understanding how a change to one parameter affects the others, as well as the underlying rheological changes, will help one to interpret SCC data more effectively. Additionally, it is important to consider if the values derived from a testing method are the result of one or more characteristics.

Slump Flow, Yield Stress, and Plastic Viscosity Relationship

Similar to the slump of conventional concrete, the slump flow of an SCC mixture increases as the yield stress decreases.[9,23] However, all SCC mixtures will have relatively low yield stress values and must be below a certain maximum value in order to self-consolidate. This is in contrast to the relatively broad range of viscosity values possible for SCC.[9] Therefore, with respect to the rheological parameters, once the slump flow level has been established, it is primarily the plastic viscosity that is manipulated in order to control performance.

V-Funnel, T50, and Viscosity Relationship

The relationship between viscosity and the V-funnel and T50 tests has been demonstrated previously.[6,9,23] In general, as the measured value for the V-funnel or T50 test increases, the plastic viscosity increases. Although these tests do not provide a true viscosity value, they do provide a means for assessing the relative change in viscosity as changes are made to SCC mixtures without the need for a concrete rheometer. These measurements are useful from a mixture development, quality control, and troubleshooting perspective.

Viscosity and Segregation Resistance

Stokes' law tells us that the frictional force on a sphere moving in a fluid will increase as the viscosity of the fluid increases. This increased frictional force slows the rate of segregation. The modeling of aggregate segregation in concrete, however, is much more complicated than this, particularly as we introduce numerous particles of varied shapes and sizes as well as the time-dependent rheological properties of concrete. However, we can say that increasing the viscosity of SCC will improve mixture stability.[10]

Slump Flow Level and Segregation Resistance

Once a mixture has been proportioned and batched, the slump flow can be intentionally increased by adding more high-range water reducer (HRWR). For a given mixture, as the slump flow increases, aggregate segregation will tend to increase. The relationship between slump flow and segregation is unique to a given set of mixture proportions and a given set of materials. The increase in segregation per unit increase in slump flow will vary by mixture. Table 3.2 shows mixture data for two sets of four SCC mixtures.[24] Each set of four mixtures was proportioned identically and run to four different slump flow levels by increasing the dosage of HRWR. The two sets differed in that one used a cement content of 435 kg/m^3 and the other used 498 kg/m^3.

Notice that as the slump flow level increases from Mixture 1 to Mixture 4 and from Mixture 5 to Mixture 8, the segregation factor per ASTM C 1610, the bleed percentage per ASTM C 262, and the VSI all increase. The degree of change, however, is different between the two sets of mixture proportions, in this case the only difference being the higher cement contents in mixtures 5–8. SCC mixtures can be proportioned to have low aggregate segregation and controlled bleed across a range of slump flow levels. It is important, however, to know this relationship for the mixture in use.

The data set in Table 3.2 also shows the inadequacy of using only a slump flow measurement to characterize an SCC mixture. To fully characterize an SCC mixture and control its production and performance, fresh property tests in addition to slump flow should be conducted.

Viscosity, Slump Flow, and Passing Ability

Aside from proportioning variables, the passing ability of an SCC mixture is influenced by the balance between slump flow and mixture viscosity.[25] In some countries, slump flow values of 500–550 mm are the norm, while in others 650–700 mm is the norm.[9] Good passing ability, as measured by the difference between the slump flow with and without the J-ring, requires a different plastic viscosity, depending upon which of these slump flow ranges is being used. A mixture with a higher slump flow value will need a relatively higher viscosity to enable appropriate passing ability. If the viscosity is too low in this mixture, separation of coarse aggregate from the mortar fraction occurs more easily, which could lead to aggregate bridging and blocking of flow. On the other hand, a mixture with a lower slump flow value will need a relatively lower viscosity to promote the required deformation of the concrete around the obstacles. If the plastic viscosity is too high in lower-slump-flow mixtures, then the passing ability as well as the filling ability may not be sufficient.

Summary

The key to SCC is control and balance of its fresh properties. A complete understanding of the science of rheology is not necessary; however, an understanding of the basic concepts is helpful. In practice, the daily control of filling ability, passing ability, and stability is what is most important, and knowing their interactions is helpful in the troubleshooting process. One should also take time in choosing the right test methods for the mixture development versus the quality control processes. During mixture development, numerous tests can be run and relationships with other tests and properties established so as to make the quality control process and testing more efficient.

Table 3.2 Slump Flow and Stability Data[24]

Mix No.	1	2	3	4	5	6	7	8
Cement (kg/m³)	431	433	437	439	493	496	496	498
Coarse Aggregate (kg/m³)	967	970	980	983	970	977	977	980
Fine Aggregate (kg/m³)	859	862	871	873	812	817	817	820
Water (kg/m³)	178	179	181	181	179	180	180	181
Polycarboxylate Ether (PCE) HRWR (ml/100 kg)	631	683	767	956	553	611	696	865
Air (%)	2.5	1.3	1.2	0.5	1.6	1.3	1.3	1.2
Slump Flow (mm)	514	610	660	743	514	584	660	737
Segregation Factor (%)	8.1	10.3	21.3	31.2	0.0	6.6	15.9	30.8
Bleed (%)	0.62	1.1	0.96	1.8	0.5	0.4	0.9	0.9
T50 (s)	3	1.5	1.4	1.2	2.5	2.8	1.5	1.1
VSI	0	0.5	1.5	3	0	0.5	1.5	3

Chapter 4

Hardened Properties of SCC

Introduction

The hardened properties of concrete are critical regardless of a mixture's workability. When SCC was introduced in North America, questions regarding hardened properties were raised based on two primary areas of concern:

1. A historical misconception that higher slump/higher fluidity concrete is of lesser "quality" than stiffer concrete.
2. Most early SCC mixtures and the initial mixture proportioning techniques for SCC used or called for using high paste and sand contents. This led to concerns about properties such as shrinkage and creep.

Is it always the case that SCC will have higher shrinkage, higher creep, or lower modulus of elasticity values than conventional concrete? Do the fresh properties of SCC in any way influence the hardened properties? The published studies on SCC hardened properties contain sometimes conflicting conclusions regarding these questions.[1] As one examines the details and the conclusions of these studies, it is helpful to review how the reported comparison is made. There are five basic ways in which SCC hardened performance is compared or presented:

1. An SCC mixture and its data are reported on their own without comparison.
2. SCC is compared with conventional concrete for a target application. In these studies, the mixture proportions for the conventional concrete and the SCC mixtures can differ significantly. When the relative basis for comparison is lower-performance conventional concrete, the difference between it and SCC is greater than when the relative comparison is higher-performance concrete.
3. SCC is compared with conventional concrete, both being proportioned to achieve a targeted design strength. Developing lower-strength SCC mixtures that exhibit the appropriate fresh properties is sometimes

difficult because to achieve these fresh properties certain limitations exist in the proportioning techniques. The difference in proportions and hardened properties will tend to diminish as the strength level targeted increases.

4. As systematic changes to SCC proportions and/or material constituents are made, the resulting changes in performance are reported. There are numerous studies describing the influence of paste volume or composition,[2] water/cement (w/c) ratio,[3] and aggregate volumes and compositions.[4,5]

5. SCC performance is compared with predictive equations from ACI, AASHTO, or other European codes. In these cases, the intention is to determine whether the typical equations used by design engineers can be applied to SCC.

The purpose of this chapter is to provide a general review of the hardened properties of SCC. This chapter will not provide predictive equations for engineering purposes, nor is it meant to replace the necessary testing required when developing mixtures for specific applications. As with conventional concrete, it is the responsibility of the individual to ensure that key hardened properties are considered as adjustments to the mixture proportions are made. A discussion of how SCC fresh properties may impact certain hardened properties is presented. In general, however, the hardened properties of SCC are driven by the materials and mixture proportions used, and changes to these follow the same rules of thumb as for conventional concrete.[4–7]

Variables Influencing Hardened Properties

Each of the following subsections will review a specific property and the proportioning variables that influence it.

Compressive Strength

SCC can be developed to achieve a wide range of compressive strength levels.[8] The compressive strength development is primarily determined by the w/c ratio and by the composition of the cementitious components. For example, SCC mixtures can be produced with any combination of cements, pozzolans, and in some cases other inert fine powders. If an SCC mixture is developed using a combination of cement and fly ash, then the compressive strength development may need to be measured out to a time past the typical 28 days,[9] as is done for conventional slump concrete. Similarly, if high early compressive strength is desired, such as for precast concrete, the replacement level of Portland cement with supplementary cementitious material will be limited.[10]

Modulus of Elasticity

The modulus of elasticity (MOE) is important, as the use of SCC has increased particularly in precast and prestressed concrete applications. Several studies have indicated that, similar to conventional slump concrete, the MOE of SCC is reduced as the mixture's paste content increases, as the strength decreases, and as the aggregate content decreases.[3,5,6,10–12] Since a wide range of compressive strength values can be developed, a wide range of MOE values can also be attained in SCC mixtures.

Both ACI and European guidelines suggest that the relationship between compressive strength and modulus of elasticity for SCC mixtures may not have the same predictive relationship as they do for conventional concrete.[8,9] The report from ACI Committee 237 recommends that for applications where the MOE is critical, it should be determined from trial mixtures.[9]

Creep

Creep of concrete is highly dependent on the mixture proportions utilized. ACI 237.R.07 says that for mixtures using the same materials and proportions run to both conventional and SCC consistency, the creep of the SCC mixture will be similar to that of the conventional slump concrete. If the relative paste content increases, or the w/c ratio increases, an increase in creep should be expected.[8,10] This same tendency exists for conventional concrete.[13]

Shrinkage and Cracking

Shrinkage of concrete is categorized as autogenous, drying, and plastic shrinkage. Plastic shrinkage occurs as the evaporation of bleed water surpasses the rate of bleed to the surface before hardening and sufficient strength development. The rate and amount of bleed can be influenced by certain mixture proportioning techniques, as can the rate of hardening and strength development. Increasing the air content and the fine material content (cement plus other powders, including sand fines) and decreasing the w/c (or water/powder) ratio will all tend to decrease the amount and rate of bleed to the surface. The inclusion of some viscosity-modifying admixtures (VMAs) can also decrease the amount and rate of bleed. The rate of hardening and strength development will be slowed by the use of retarding admixtures, by replacing cement with supplementary cementitous materials, or by using very high dosages of certain high-range water reducers (HRWRs). SCC mixtures proportioned with high powder contents, low water/powder ratios, and/or high dosages of certain VMAs can be more susceptible to plastic shrinkage cracking, given the appropriate environmental conditions. In these cases, care should be taken to minimize surface evaporation at early ages.

Drying shrinkage is primarily impacted by the characteristics of the constituent materials, as well as by the mixture proportions used.[8–10,14] Table 4.1 presents drying shrinkage data for 14 concrete mixtures.[15] The evaluation consisted of two series of seven mixtures each. Each series used a design water content of either 170 or 192 kg/m^3 and compared a reference conventional slump concrete mixture with six SCC mixtures. The SCC fresh properties were achieved through the use of an HRWR admixture and a VMA. The purpose of the evaluation was to compare drying shrinkage, as measured according to ASTM C 157, of conventional concrete with that of SCC mixtures and to investigate the influence of SCC proportioning changes on drying shrinkage. The six SCC mixtures in each series were proportioned with a cement content of 430 or 505 kg/m^3 and a low, medium, or high coarse aggregate content. The cement used was a Type I Portland cement and the coarse aggregate was a crushed limestone with a maximum size of 25 mm.

Mixtures 1 and 8 are the reference conventional slump concrete mixtures at a cement content of 430 kg/m^3. Figure 4.1 shows the 28-day shrinkage percentage as a function of coarse aggregate content per cubic meter. Two graphs are included, one for each water content. The shrinkage of SCC follows the same rules of thumb as for conventional concrete. Shrinkage increases as water content increases and as coarse aggregate content decreases. One should also notice that the conventional slump mixtures with similar proportions produce similar shrinkage to the SCC mixtures.

Shrinkage of concrete is important, as it relates to cracking. ASTM C 1581 was developed as a method for assessing the cracking potential of concrete as it shrinks under restraint. See and Attiogbe[16] evaluated the shrinkage and cracking potential of several SCC mixtures versus conventional concrete. In their study, two sets of mixture proportions were developed, one using rounded and one using angular coarse aggregates. Both sets of proportions were run to a 150–200 mm slump and a 675–700 mm slump flow by increasing the HRWR only. With proportions and materials identical and the only difference being the workability level, they found that the 28-day shrinkage and the time to cracking were virtually identical for each set of materials and mixture proportions and that no influence from workability level was seen.[16] Lowering the sand/aggregate ratio, minimizing the paste volume, or including fibers and shrinkage-reducing admixtures can reduce shrinkage and cracking potential of SCC mixtures.[16,17]

Bond to Steel and Prestressing Strand

The bond to steel of conventional concrete can be impaired if the mixtures are over-vibrated. If a suitably fluid and stable SCC mixture is used instead, this variable in construction can be removed.[18] The enhanced fluidity of SCC has been shown to positively impact its bond to reinforcing steel as long as the SCC mixture is stable.[9] As an SCC mixture becomes less stable, the bond

Table 4.1 Shrinkage Comparison of Conventional Slump and SCC Mixtures

Mix No.	1	2	3	4	5	6	7	8	9	10	11	12	13	14
Cement (kg/m³)	430	437	430	426	509	505	500	441	433	433	426	512	512	508
Coarse Aggregate (kg/m³)	1067	1160	950	754	1110	918	728	1095	1110	924	728	1079	899	713
Fine Aggregate (kg/m³)	782	718	882	1050	696	862	1025	733	690	862	1018	671	862	997
Water (kg/m³)	170	173	170	168	171	170	168	197	195	195	191	196	196	194
Paste (%)	31	31	31	30	33	33	33	34	33	33	33	36	36	36
Sand/Aggregate Ratio	0.44	0.40	0.50	0.60	0.40	0.50	0.60	0.42	0.40	0.50	0.60	0.40	0.50	0.60
PCE HRWR (ml/100 kg)	325	637	605	735	559	559	605	176	332	410	455	423	449	468
VMA (ml/100 kg)	0	195	195	130	98	49	0	0	195	195	130	98	49	0
Slump (mm)	178							178						
Slump Flow (mm)		641	610	641	610	622	610		641	616	610	616	622	610
VSI	0	1	1	1	0	1	1	0	2	1	1	1	0	0
Air (%)	1.3	0.3	1.3	1.4	0.7	1.6	2.2	0.9	1.6	1.4	2.4	0.9	1.6	2.5
28-Day Shrinkage (%)	-0.027	-0.022	-0.024	-0.028	-0.018	-0.021	-0.028	-0.03	-0.032	-0.032	-0.034	-0.028	-0.035	-0.038

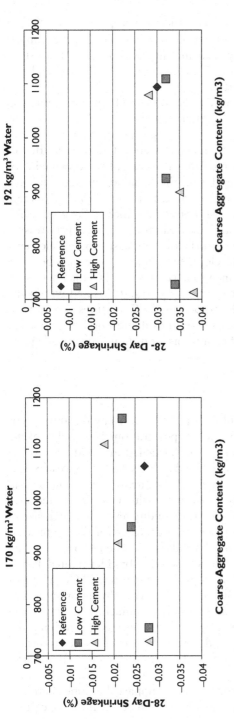

Figure 4.1 The 28-Day SCC Drying Shrinkage as a Function of Coarse Aggregate Content.

can be compromised. This is particularly true with respect to the top-bar effect in taller elements, where migrating bleed water can become trapped underneath the upper bars, thereby reducing the bond of concrete to steel. This effect was shown to increase as mixture settlement increased as determined from the surface settlement test.[9]

A number of studies have also investigated SCC and bond to prestressing strand. Burgueno and Haq[19] reported different trends of pull-out strengths and transfer, flexural, and development lengths, indicating that the bond performance and the bond-related parameters were different for different SCC mixes. Hegger et al.[20] commented that the general bond behavior of pretensioned strands in SCC is similar to the behavior in high-strength concrete. However, the bond strength of SCC is sensitive to the concrete mixture used. In April 2007, the Precast Pretressed Concrete Institute issued a technical bulletin based on industry research indicating that prestressing strands located near the as-cast upper surfaces of precast concrete elements may exhibit greater transfer and development lengths. There may be some similarity here to the top-bar effect mentioned earlier. All PCI members were advised to examine strand slip for prestressing strands located within the upper region (as-cast) of their products. Because SCC mixtures vary in their materials and proportions, producers should evaluate bond performance based on their materials and sets of conditions, and should ensure adequate mixture stability.

Durability

SCC can be proportioned to provide adequate freeze/thaw durability, salt scaling resistance, and resistance to chloride penetration. The quality of the aggregates utilized and the quality of the hardened air void system are important to the salt scaling and freeze/thaw durability performance of SCC.[9] The air system in concrete is created by the folding action during the mixing process. Many things influence the ease with which air is created and stabilized in a mixture, including the mixture fluidity level, the batch size in relation to the mixer size, mixture proportions, fineness of the aggregates, the cementitious system, and chemical admixtures. The action of an air-entraining admixture (AEA) in concrete is to stabilize bubbles created during mixing. Table 4.2 shows the spacing factor of air-entrained SCC mixtures with target slump flow of 610–660 mm, made with three commercially available AEAs.[21] Air void analysis samples were manufactured in two ways: by pouring and filling the cylinders only and by pouring and rodding the cylinders. The data show that measured spacing factors and specific surface areas improve when the specimens are rodded versus when they are poured only.

Casting a cylinder in the laboratory does not impart the same energy as the placement process during production. One should be careful when casting specimens for air void analysis, as the filling process can entrap voids and cause the air void system to appear worse than it is.

Concrete permeability increases with w/c ratio and decreases with the inclusion of materials such as silica fume. Table 4.3 shows rapid chloride permeability[22] test results for SCC mixtures made both with and without silica fume.[23] Notice the improvement in 28- and 56-day ratings for the mixtures as silica fume is added. This same trend exists for conventional concrete.

Table 4.2 Hardened Air Void Parameters for Poured versus Rodded Specimens

Mix No.	1	2	3
Cement (kg/m³)	504	504	504
Coarse Aggregate (kg/m³)	769	769	769
Fine Aggregate (kg/m³)	872	872	872
Water (kg/m³)	170	170	170
PCE HRWR (ml/100 kg)	592	527	592
AEA Product	1	2	3
AEA (ml/100 kg)	20	13	13
Air (%)	6.9	6.5	6.9
VSI	0	0	1
Air Void System, Poured			
Hardened Air (%)	5.9	8.8	5.5
Specific Surface (1/mm)	26	18	30
Spacing Factor (mm)	0.203	0.203	0.178
Air Void System, Rodded			
Hardened Air (%)	4.6	6	3.9
Specific Surface (1/mm)	34	27	50
Spacing Factor (mm)	0.152	0.178	0.127

Table 4.3 Rapid Chloride Permeability with and without Silica Fume

Mix No.	1	2	3
Cement (kg/m³)	458	456	446
Silica Fume (kg/m³)	0	27	54
Coarse Aggregate (kg/m³)	797	780	746
Fine Aggregate (kg/m³)	912	891	855
Water (kg/m³)	173	172	168
HRWR (ml/100 kg)	510	870	1435
AEA (ml/100 kg)	50	50	98
Air (%)	5.1	5.3	7
Slump Flow (mm)	600	685	685
V-Funnel Time (s)	6.1	6.0	5.9
ASTM C 1202 RCP, Coulombs			
28-day	4362	1832	753
56-day	3216	1317	365

The mixture proportions, primarily the quality and density of the cement paste, play a significant role in the durability of SCC mixtures. Developing an SCC mixture with compressive strength equivalent to a conventional concrete mixture is not sufficient to reach a similar durability behavior. When comparing durability properties based on equal cement content and equal w/c ratio, however, the durability of SCC is as good as that of conventional concrete.[23]

Mixture Fluidity, Stability and Hardened Properties of SCC

It stands to reason that the hardened properties are heavily influenced by the mixture proportions, but the question is whether the properties that differentiate SCC from conventional slump concrete will have an influence. What influence, if any then, does fluidity and mixture stability have on the hardened properties of SCC?

Fluidity Impact on Compressive Strength

The compressive strength for an SCC mixture can be higher than that of a conventional slump mixture at the same w/c ratio.[8,9] Table 4.4 shows data from a laboratory investigation designed to study the impact of fluidity level on compressive strength.[25] In this evaluation, four large batches of concrete were produced. From each batch, four sets of compressive strength cylinders were cast at four fluidity levels. Each mixture was initially dosed with HRWR to achieve a slump of 125–150 mm, at which point a set of compressive strength cylinders was made. Additional HRWR was added to the remaining concrete, the slump was increased to 200–225 mm, and another set of strength cylinders was made. This process was repeated twice more, achieving slump flow levels of 500–550 and 600–650 mm. Each time, additional cylinders were made for compressive strength testing. In this evaluation, Mixtures 1 and 2 and Mixtures 3 and 4 used 10 and 25 mm top size coarse aggregate, respectively.

Figure 4.2 shows the data from Table 4.2 graphically. These data support the observations of other researchers that the compressive strength of SCC mixtures with the same w/c ratio can be higher than that of a conventional slump mixture.

Mixture Fluidity and Viscosity Impact on Surface Finish

To study the impact of fluidity and viscosity levels on the formed surface finish, an experiment was designed by changing water content, VMA dosage, and slump flow level.[26] Each variable was tested at two levels: water content 180–197 kg/m^3, VMA dosage 0–260 ml/100 kg, and slump flow level 510–560 and 650–700 mm. The T50 time was measured to provide a relative

Table 4.4 Concrete Mixture Fluidity and 28-Day Compressive Strength

Mix No.	1				2				3				4			
Cement (kg/m³)	490	490	490	490	484	487	490	490	490	492	490	489	488	493	493	484
Coarse Aggregate (kg/m³)	817	817	817	817	783	788	792	792	817	920	817	815	790	797	797	783
Fine Aggregate (kg/m³)	952	952	952	951	912	917	92	923	952	955	952	949	920	928	928	912
Water (kg/m³)	167	167	167	167	184	185	186	186	167	167	167	166	186	187	187	184
HRWR (ml/100 kg)	652	717	848	913	261	326	456	521	652	848	913	978	456	618	683	813
Slump (mm)	145	220			145	240			115	215			120	190		
Slump Flow (mm)			590	690			575	680			550	690			525	675
Air (%)	3	3.2	2.6	1.9	3.4	3.2	2.6	2.1	3	2.5	2	2.2	3.2	2.1	2.4	1.4
Compressive Strength 28-Day (MPa)	71	70.6	76.2	73.7	60.3	58.4	64.9	68.5	68.5	76	78.4	79	57.3	65.8	67.9	75.5
Relative to Lowest Slump	100%	99%	107%	104%	100%	97%	108%	114%	100%	111%	114%	115%	100%	115%	118%	132%

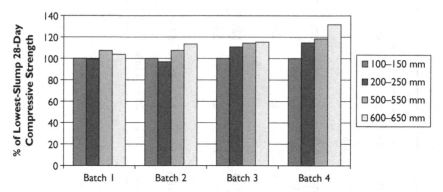

Figure 4.2 The 28-Day Compressive Strength by Slump/Slump Flow Level.

indication of the mixtures' viscosities, and the surface finish was evaluated by visually assessing the cast surface of an H-shaped form (Figure 4.3). Each SCC mixture was poured from a bucket into one side of the form and allowed to flow through the form, across the horizontal section, and fill the opposite vertical section. Although not timed specifically, the pouring rate was fairly high and held constant. After hardening, the forms were stripped and visually examined for surface imperfections. A surface defect rating of 0–3 was given to each, with 0 indicating no imperfections and 3 meaning significant imperfections. Table 4.5 shows the data generated during this study. In general, the group with higher water content had better surface finish.

Figure 4.4 shows the relationship between slump flow, T50, and surface defect rating. In this study, the viscosity was shown to have the greatest impact, while the slump flow level had no real impact on the formed surface finish.

This study evaluated the finish of a flat surface only. When evaluating the surface finish of more intricately detailed elements, the slump flow will become more important.

Mixture Stability Impact on Compressive Strength and Modulus of Elasticity

There is limited information available in the literature discussing the impact of mixture stability on compressive strength values. Some studies indicate that for identical mixture proportions, as the segregation increased, there was no real impact on the compressive strength or the modulus of elasticity.[27] Table 4.6 shows an additional set of data comparing segregation percentage with the 24-hour compressive strength and modulus of elasticity of six mixtures.[28] Two sets of three mixtures were run and all six used the same mixture proportions. The only differences were that one set of three mixtures

Figure 4.3 H-Shaped Forms to Assess Surface Finish.

Table 4.5 Surface Defects as a Function of Slump Flow and Viscosity

Mix No.	1	2	3	4	5	6	7	8
Cement (kg/m³)	430	430	430	430	438	443	438	443
Coarse Aggregate (kg/m³)	919	919	919	919	938	947	938	947
Fine Aggregate (kg/m³)	887	887	887	887	905	914	905	914
Water (kg/m³)	197	197	197	197	181	183	181	183
HRWR (ml/100 kg)	429	592	637	800	592	696	696	897
VMA (ml/100 kg)	0	0	260	260	0	0	260	260
Slump Flow (mm)	510	685	560	700	560	660	535	660
T50 (s)	1.5	1	3.7	1.8	2.8	1.6	4.2	3.4
Surface Defect Rating (0–3)	1	1	3	2	2	2	3	3

did not use a VMA and one set did. In addition, each set of three mixtures was run to three workability levels: a conventional slump and a low- and high-slump flow. Comparing the compressive strengths of the conventional slump mixtures to the SCC mixtures in Mixtures 1–3 and 4–6, we see no loss in compressive strength or in modulus of elasticity as the segregation increases.

In this set of evaluations, the stability of the mixture had no effect on the hardened properties measured. Other studies measured the relative sorptivity and rapid chloride permeability of stable versus unstable SCC mixtures. As

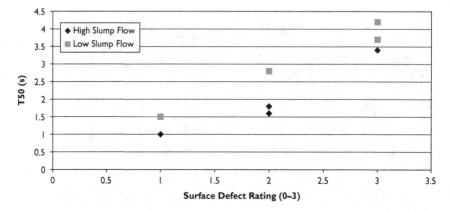

Figure 4.4 Slump Flow, T50, and Formed Surface Defects.

Table 4.6 Mixture Stability and Compressive Strength

Mix No.	1	2	3	4	5	6
Cement (kg/m³)	426	421	432	423	423	438
Fine Aggregate (kg/m³)	920	910	935	913	915	946
Coarse Aggregate (kg/m³)	881	871	895	874	875	904
Water (kg/m³)	176	174	179	175	175	181
HRWR (ml/100 kg)	190	495	665	230	560	930
VMA (ml/100 kg)	No	No	No	Yes	Yes	Yes
Slump (mm)	165	NA	NA	160	NA	NA
Slump Flow (mm)	NA	600	735	NA	620	710
T50 (s)	NA	2.03	2.71	NA	2.81	2.15
Air (%)	2.4	4.2	0.5	2.9	3.2	0.4
Unit Weight (kg/m³)	2405	2379	2444	2385	2391	2470
VSI	0	1	3	0	1	3
Segregation (%)	0	10.2	17.5	0	0	11.7
24 h Compressive Strength (MPa)	24.3	26.7	25.2	20.4	25.2	21.4
24 h Modulus of Elasticity (GPa)	22.75	22.75	22.82	22.30	22.18	21.30

NA: not available.

the mixture stability decreased, the relative sorptivity and rapid chloride permeability between samples taken from the top and bottom of a laboratory cast column increased, indicating poorer performance on the top of sections cast with unstable SCC.[28]

Mixture Stability and the Air Void System

As SCC mixtures become unstable and segregation occurs, entrained and entrapped air bubbles will rise along with any free water as the heavier particles sink. This instability will compromise the integrity of the air void system and as such can impact the durability of an SCC mixture. Table 4.7 shows the spacing factor of a mixture run to three fluidity levels and three corresponding stability levels.[29] In this evaluation, a single 200 mm by 400 mm cylinder was cast from each mixture for air void analysis. The top and bottom quarters were cut from each cylinder and an analysis was performed on the top and bottom of each section, so that four measurements of the air void system were performed on each specimen. The data in the table show that as a mixture becomes less stable, the air void system is negatively impacted. As the mixture becomes highly unstable (VSI = 3), not only does the spacing factor increase, but within a single specimen it becomes highly variable.

Summary

More than 1000 papers have been written on SCC and many of these investigations have studied particular hardened properties. Some have concluded that SCC's hardened properties are not equivalent to those of conventional slump concrete, while others have concluded that the properties are similar to or better than those of conventional slump concrete. Either way, it is obviously impossible to categorize SCC as always worse or always

Table 4.7 Stability Influence on the Hardened Air Void Spacing Factor

Mix No.	1	2	3
Cement (kg/m³)	436	436	439
Coarse Aggregate (kg/m³)	903	903	910
Fine Aggregate (kg/m³)	814	814	820
Water (kg/m³)	180	180	182
PCE HRWR (ml/100 kg)	319	417	640
AEA (ml/100 kg)	16	16	58
Slump Flow (mm)	572	654	737
T50 (s)	2.3	2.3	3.0
Air (%)	6.0	5.3	5.0
VSI	0	1	3
Spacing Factor (mm)			
Top Quarter, Top	0.290	0.335	0.447
Top Quarter, Bottom	0.287	0.330	0.687
Bottom Quarter, Top	NA	0.307	0.479
Bottom Quarter, Bottom	0.291	0.339	0.359

NA: not available.

better. The conclusion is that SCC's hardened properties are primarily dictated by the mixture proportions and constituent materials used. If the proportions differ significantly from the conventional slump mixture with which it is compared, then one should expect a difference in performance. Having a mixture that is stable is necessary for this statement to be true: if a mixture becomes less stable, certain properties can be negatively impacted, while others are not as strongly influenced. The fluidity level of SCC negatively impacts performance only in so much as it affects stability. It should be noted that highly fluid and stable SCC mixtures with excellent hardened properties can be and are produced regularly.

Chapter 5

Powders and Aggregates

Introduction

It is the combined properties of the constituent materials that drive SCC performance. Because material characteristics and availability differ from location to location, it is nearly impossible to blindly take a set of SCC mixture proportions from one geographic location to another and achieve identical performance. Clearly, the high-performance properties of SCC cannot be achieved through a prescribed mix design. Prevalent aggregates in one area may be crushed dolomitic limestone, while in another rounded river gravel is available, or while slag cement is available in one place, Class C fly ash may be available in another. The local concrete producer must understand the differences in locally available materials and be free to adjust proportions appropriately to achieve the right performance. This chapter will focus primarily on the physical characteristics of the constituent materials, as the chemical reactivity, particularly of the powders, is beyond the scope of this book.

Powders

A powder is a material with a particle size of 125 μm or smaller that is intentionally added to an SCC mixture for the intention of increasing paste volume and/or improving the concrete rheology. Powders are extremely important to SCC mixtures. In combination with water and air, they compose the paste fraction of concrete, which is the portion of the mixture that provides mobility. A number of powders, useful to SCC, exist and can be classified into three main categories: cements, pozzolans, and fillers. Cement and pozzolans are well known and include those materials governed by current industry standards such as ASTM C 150, C 1157, and C 618.[1-3] Fillers, on the other hand, are not yet classified according to any concrete industry standard, although the author is aware of efforts being made in groups such as ASTM and ACI to define and classify filler materials.

Powder characteristics such as fineness, particle size distribution, particle shape, composition, and reactivity, and specific gravity affect the fresh properties of SCC.[4-6] The fresh properties most influenced by the powder composition include initial slump flow, which affects water or high-range water-reducer (HRWR) demand, workability retention, viscosity, and bleed. The water sensitivity of a powder is also important, as it will influence the overall robustness of the SCC mixture itself. A process for categorizing powders based on water sensitivity has been developed and can be useful to concrete producers who have numerous powders available.[6,7]

Most powders have a lower specific gravity than Portland cement. When a lower-specific-gravity powder is used as a cement replacement, this is typically done by mass, which leads to an increase in paste volume. As will be discussed in more detail in Chapter 8, the paste volume is a critical component of SCC mixture proportioning strategies. A substantial increase in paste volume is possible, depending upon the specific gravity of the material and the replacement percentage. For example, assuming an SCC mixture with 450 kg/m³ of cement, if 25% of the cement is replaced by a powder with a specific gravity of 2.4, a greater than 1% increase in paste volume would occur. Numerous papers have been written on how particular powders influence SCC performance. When reviewing data from these studies, knowing how the experiment was designed is critical. Typically, powders will be directly compared with one another or with a straight cement mixture. In either case, one must know whether the experiment was designed to maintain a constant paste volume or a constant mass replacement. An increase in paste volume will enhance the fresh SCC properties and if not considered during the data analysis phase could lead to erroneous conclusions regarding material comparisons.

Cements

The ability to test Portland cement is not found in the average quality control laboratory of most concrete producers. That does not mean, however, that a producer should ignore the understanding of cement characteristics, particularly if one uses multiple cements interchangeably. Which characteristics of Portland cement should one consider when evaluating the fresh properties of SCC?

Particle Shape

The particle shape of Portland cement is angular. If replaced by another powder with a round particle shape, then the fresh properties of the mixture will be influenced.

Fineness

The fineness of cement gives an indication of the surface area per unit mass. All other things being equal, as the fineness of cement increases, the HRWR demand or water demand may increase in order to achieve a certain slump flow. Greater fineness can increase the viscosity of the cement paste as well as the rate of cement hydration, thereby shortening the setting time and increasing the rate of strength development.[8] A faster rate of hydration can also shorten the workability retention time of an SCC mixture.

Water Demand

The water demand of cement can differ by source and can impact the admixture dosages required for achieving SCC properties. Besides the fineness and surface area, it has been reported that the sulfate type and content, alkali level, and particle size distribution can also affect the water demand of Portland cement.[9] ASTM C 187, Normal Consistency, can be used to determine the water demand differences between various cements.[10]

Pozzolans

Pozzolans such as fly ash, silica fume, and slag cement are used extensively in conventional concrete to economize mixtures, lower the heat of hydration for mass concrete, reduce chloride permeability, and inhibit alkali–silica reactivity, among other things. In recent years, they have been used to enhance the sustainability and reduce the carbon footprint of concrete by reducing the Portland cement content per cubic meter or cubic yard. These materials can be and are used in SCC for the same purposes.

Fly Ash

In conventional concrete, for durability reasons, fly ash is limited to 25% replacement by weight of cement for the most severe exposure class per ACI 318.[11] The specific gravity of fly ash can have a broad range from 1.9 to 2.8,[12] which will increase the paste volume and enhance the fresh properties of the mixture. The average particle size of fly ash is similar to that of Portland cement, but the round particle shape of fly ash enhances flowability when used as a replacement for the more angular particle shape of cement. This can permit a reduction in water content or HRWR dosage. If water is reduced, bleed as well as paste volume will also be reduced. Studies in mortar have shown that loss on ignition (LOI), which is used as an indicator of carbon content in fly ash, influences flowability. A higher LOI adversely affects the initial flowability of SCC mortar and therefore may result in higher water demand and/or higher HRWR requirement to achieve a target slump flow.[13]

Slag

Similar to fly ash, for durability reasons, ACI 318 limits slag to 50% replacement of Portland cement for the most severe exposure class. Unlike fly ash, the particle shape of slag cement is similar to that of Portland cement; therefore, if replacing Portland cement with slag cement, no significant impact from particle shape will likely be seen.

Silica Fume

ACI 318 limits the amount of silica fume replacement to 10% of the weight of Portland cement. The silica fume particle size is much smaller than that of most other powders available. Either when replacing cement or when used as an additive, silica fume may result in an increased demand for water or for HRWR in order to maintain a given slump flow. Some studies suggest that when added in small amounts, silica fume can reduce the viscosity of a mixture but that at some point, as more silica fume is added, this behavior reverses and the viscosity is increased.[14] In this context, the addition of silica fume in small amounts, up to approximately 4%, is one of the few ways of decreasing the viscosity of an SCC mixture when the water/powder (w/p) ratio or the paste volume cannot be increased enough.[15]

Table 5.1 shows the basic physical properties of Portland cement, fly ash, slag cement, and silica fume. It can be seen that the average particle sizes and the surface areas of Portland cement, fly ash, and slag cement are similar, but that silica fume is quite different. Knowing this is helpful when considering the overall solid particle size distribution during the materials selection and mixture proportioning process.

Other Powders

Several other, nonstandardized, powders have been discussed in the literature. They can be classified as either waste or processed materials. An example of a waste material is crusher dust, a by-product of aggregate production, whose mineralogy is dependent upon the parent material. This material can be variable from source to source and sometimes within a source owing to a

Table 5.1 Physical Characteristics of Portland Cement, Fly Ash, Slag Cement, and Silica Fume[12]

	Specific Gravity	Average Particle Size (μm)	Surface Area (m^2/kg)	Particle Shape
Portland Cement	3.15	15	300–600	Angular
Fly Ash	1.9–2.8	<20	300–500	Round
Slag Cement	2.85–2.95	<45	400–600	Angular and rough
Silica Fume	2.2–2.5	0.1	20,000	Round

lack of quality control or standardization. Use of this type of material is possible for SCC, but should be preceded by a thorough testing program to investigate its impact on the fresh and hardened properties. If one is to consider using a waste material, some minimum standard, with the supplier addressing material consistency, should be agreed upon. Processed materials such as ground or milled limestone or precipitated calcium carbonate will be more consistent within a given source, as these materials are manufactured for a given purpose. It is also possible to obtain very specific products, in terms of physical characteristics, from the manufacturers of these processed materials. Similar to cements and pozzolans, these powders can influence the fresh properties of SCC based on particle shape, fineness, and particle size distribution. The effects will typically be on water or HRWR demand, overall rheology, and water sensitivity. Depending on the source, these other powder materials have the potential to economize SCC mixtures.

Aggregates

Aggregates comprise between 60% and 80% of the total volume of a concrete mixture.[16] In an SCC mixture, one relies on the paste fraction to promote fluidity and to provide the fresh property benefits of SCC. On their own, aggregate particles do not enhance the flowability of an SCC mixture. Interparticle friction between aggregate particles consumes the flowing energy of the paste during placement, thereby reducing rather than enhancing flowability. This becomes especially relevant as the concrete is required to flow through restricted sections.[17] In light of this, SCC mixtures need to be proportioned so as to limit the negative effects of aggregates on the fresh properties while remembering that it is also necessary to maintain balance between SCC's fresh and hardened properties. A low volume of aggregate may cause increased shrinkage and cracking potential, while too high a volume will inhibit flow and self-consolidation. To visualize this relationship between paste content and aggregate interparticle friction, consider a mixture in which, regardless of how much HRWR is added, the slump flow will not increase and one sees a large pile of aggregate in the middle of the slump-flow patty or discolored bleed water coming to the surface, such as is seen in Figure 5.1. In this situation, it is likely that the mixture simply has insufficient paste to facilitate movement of the aggregates. The addition of larger amounts of HRWR only makes the paste more fluid and induces segregation. Increasing the amount of chemical admixture cannot always overcome poor mixture proportioning.

The separation of aggregates during flow through restricted sections (passing ability) or by settlement is one of the primary stability concerns during the development of SCC mixtures. Passing ability is a flowing (dynamic) segregation problem governed by proportioning as well as how the paste and aggregate characteristics interact during flow. Static segregation is also driven

Figure 5.1 Slump Flow Patty with Aggregate Concentration in the Center. (Courtesy of BASF Construction Chemicals, Photographer Hal Stata.)

by the ability of the aggregates to move and settle through the paste. These stability issues can be controlled through appropriate material selection and mixture proportioning. In either case, however, when it comes to developing SCC mixtures with enhanced fresh properties, one is making choices in an attempt to limit the negative influence of the aggregates. Therefore, clearly understanding how the aggregates impact fresh properties is critical.

Coarse Aggregate Properties

Packing Density

The packing density of the coarse aggregate determines the void space between solid particles that needs to be filled with mortar. The void content is one consideration during the SCC mixture proportioning process. A more optimum packing density has also been shown to decrease the viscosity in SCC mixtures.[18] Packing density is influenced by both the particle size distribution and the particle shape.

Maximum Size

SCC mixtures have been produced with maximum size of coarse aggregates ranging from 10 to 40 mm.[19] The larger the maximum size of aggregate used, the poorer the passing ability for a given dimension through which the concrete is required to pass.[20] It should be noted that the relationship between maximum size of aggregate and passing ability is significantly affected by the proportion of coarse aggregate contained in the mixture, particularly the proportion of the larger sizes. This is why the initial proportioning method by Okamura recommended fixing the coarse aggregate content at 50% of the mixture's solid volume.[21] Larger aggregate particles will settle and segregate from the mixture more easily than smaller particles. In combination with the particle size distribution (or gradation), the maximum size will impact the total aggregate surface area, affecting paste requirement during proportioning.

Gradation

The particle size distribution influences the packing density of the aggregate. A broader range of particle sizes improves the packing density. On the other hand, a more monosized aggregate distribution decreases packing density and increases the likelihood of interparticle friction.

Particle Shape

Aggregate shape is generally classified by how angular or rounded the particles are. The overall shape can be categorized as cubical, spherical, flat, or elongated. Aggregate roundness is described as angular, subangular, subrounded, rounded, or well-rounded. Figure 5.2 shows examples of rounded versus angular and cubical versus elongated particle shapes of coarse aggregates. The particle shape has an impact on the packing density, with more-rounded aggregates having a relatively higher packing density than angular ones. Aside from influencing the required paste content, the particle shape will influence the mobility of the aggregate as the mixture flows. More-rounded and dimensionally equal particles will tumble and rotate more freely during placement. It has been shown that, at identical volumes, rounded aggregates will produce a lower-viscosity mixture than angular particles in the same mix design.[22] Particle shape will also influence the manner in which an aggregate particle passes through a restricted section (passing ability).

Specific Gravity

The specific gravity of the coarse aggregate, especially lightweight coarse aggregate, will impact segregation potential. The large particles in lightweight aggregate tend to have a lower density than the smaller particles.[23] These

Figure 5.2 Coarse Aggregate Particle Shape.

particles will have a tendency to float to the surface more readily rather than sink, like normal-weight aggregate when the mixture is unstable.

Fine Aggregate Properties

Mineralogy

In certain geographic locations, certain clays are more prevalent than in others. These clays are typically found in the finer particle size range and can cause an increase in the water demand and/or HRWR requirement to achieve a certain slump flow level.

Fineness Modulus

The fineness modulus (FM) gives an indication of the coarseness of the fine aggregate. Typically, concrete mixtures proportioned with higher-FM fine aggregates will tend to bleed more (see Table 5.4 below). In SCC, the bleed tendency will also be a function of the water content and the total powder content, which includes the cement and pozzolans. The FM does not provide sufficient information for use in proportioning SCC mixtures. However, it can be used as an indication of fluctuations in material quality.

Gradation

Similar to coarse aggregate, the particle size distribution of the fine aggregate will influence the packing density and therefore the paste required to fill the void space between fine aggregate particles and promote flow.

All other things being equal, the amount of finer material in the aggregates has a direct impact on the bleed percentage in SCC mixtures. Table 5.2 shows the gradation and fineness modulus of two fine aggregates: FA 1 and FA 2. Both of these fine aggregates were run in a series of 24 SCC mixtures to evaluate the impact of the fine aggregates on bleed. The materials used included a standard ASTM C 150 Type I cement, and three limestone coarse aggregates from the same source but with different gradations. Table 5.3 shows the gradation information for the three coarse aggregates.

The experiment was designed to produce mixtures with a range of stability levels in order to assess the impact of the aggregates. To vary the fine material content, two cement factors and two sand-to-total-aggregate ratios were used. Table 5.4 shows the mixture proportions and the fresh property data from the experiment.[24]

Table 5.2 Fine Aggregate Characteristics

	Cumulative % Passing		Individual % Retained	
	FA 1	FA 2	FA 1	FA 2
4.75 mm (#4)	99.8	100	0.2	0
2.36 mm (#8)	88.1	87.9	11.7	12.1
1.18 mm (#16)	70.7	66.5	17.4	21.4
600 μm (#30)	49.5	38.3	21.2	28.2
300 μm (#50)	19.8	12	29.7	26.3
150 μm (#100)	4.4	1.8	15.4	10.2
75 μm (#200)	1.9	0.5	2.5	1.3
Fineness Modulus, FM	2.68	2.94		

Table 5.3 Coarse Aggregate Gradations

	Cumulative % Passing			Individual % Retained		
	CA 1	CA 2	CA 3	CA 1	CA 2	CA 3
50 mm (2 in.)	100	100	100	0	0	0
37.5 mm (1½ in.)	100	100	100	0	0	0
25 mm (1 in.)	99	100	98.4	1	0	1.6
19 mm (¾ in.)	92.2	100	82.8	6.8	0	15.6
12.5 mm (½ in.)	65.4	100	33.8	26.8	0	49
9.5 mm (⅜ in.)	45.8	77.4	21.8	19.6	22.6	12
4.75 mm (#4)	2.4	3	3.2	43.4	74.4	18.6
2.36 mm (#8)	1.4	0.1	3	1	2.9	0.2
1.18 mm (#16)	1.4	0.1	3	0	0	0

Table 5.4 Mixture Proportions, Fresh Properties, and Fine Material

Mixture No.	1	2	3	4	5	6	7	8	9	10	11	12
Cement (kg/m³)	426	428	432	434	436	439	430	432	434	436	436	441
Coarse Aggregate (kg/m³)	808	812	820	824	828	833	964	969	974	978	978	989
CA 1, 2, 3	2	1	2	3	1	3	1	2	2	1	3	3
Fine Aggregate (kg/m³)	973	977	1003	992	1013	1018	843	847	865	869	856	879
FA Type 1, 2	2	1	2	2	1	1	2	2	1	1	2	1
Water (kg/m³)	176	177	178	180	181	181	178	178	180	181	180	183
PCE HRWR (ml/100 kg)	1294	683	1801	624	1515	1430	735	819	1066	1021	696	904
Air (%)	1.3	2.3	1.6	1.7	1.5	1.4	1.4	1.8	1.7	1.6	0.9	1.2
Slump Flow (mm)	686	686	699	686	711	699	699	686	711	692	711	686
Bleed (%)	30.0	6.6	17.6	23.8	0.1	0.2	29.9	26.7	2.1	0.4	46.5	0.4
Column Segregation (%)	4.0	26.8	12.7	32.7	12.7	18.7	25.3	14.5	15.1	27.9	20.7	32.8
T50 (s)	1.7	1.8	3.5	1.4	2.5	2.8	2.8	1.7	2.3	2.0	1.6	2.0
VSI	3.0	3.0	2.5	2.5	2.0	2.0	2.5	3.0	2.0	1.5	3.0	1.0
FA Material Passing 300 µm Sieve + Cement (kg)	542	621	553	553	637	640	531	534	605	609	539	615

Table 5.4 Continued

Mixture No.	13	14	15	16	17	18	19	20	21	22	23	24
Cement (kg/m³)	487	494	496	496	496	496	494	494	494	496	499	502
Coarse Aggregate (kg/m³)	812	824	828	828	828	828	974	974	974	978	984	989
CA 1, 2, 3	2	2	3	–	3	–	2	2	3	–	–	3
Fine Aggregate (kg/m³)	929	958	962	962	948	948	814	802	802	806	822	827
FA Type 1, 2	2	–	–	–	2	2	–	2	2	2	–	–
Water (kg/m³)	177	180	181	181	181	181	180	180	180	180	181	183
PCE HRWR (ml/100 kg)	936	1619	1222	1294	683	865	1040	754	650	650	1008	897
Air (%)	1.9	1.1	1.5	1.8	1.6	1.7	1.6	1.4	–	1.1	–	1.3
Slump Flow (mm)	699	718	711	686	686	686	711	711	711	699	711	686
Bleed (%)	5.1	0.1	0.1	0.0	0.8	4.4	0.3	8.6	14.3	5.7	0.2	0.4
Column Segregation (%)	23.1	22.2	6.3	11.9	25.6	16.7	22.6	30.2	34.6	13.3	15.2	19.3
T50 (s)	2.1	2.9	2.7	3.2	2.9	1.9	2.8	2.1	1.4	3.1	2.6	2.2
VSI	3.0	3.0	1.0	1.0	2.5	2.0	1.0	2.5	2.5	1.0	2.0	1.5
FA Material Passing 300 μm Sieve + Cement (kg)	598	684	687	687	610	610	655	590	590	593	662	665

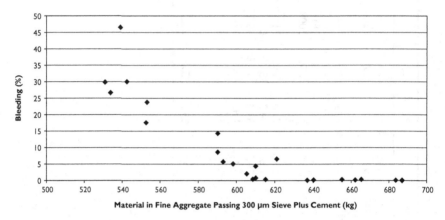

Figure 5.3 Bleeding as a Function of Total Material Passing the 300 μm Sieve, Including Cement.

In analyzing the data, it was found that bleeding is strongly correlated not only to the powder material (cement, pozzolan, etc.), but also to the total amount of material finer than the 300 μm sieve in the concrete mixture coming from both the powder and the aggregates. Figure 5.3 shows this relationship.

It should be noted that if the water content of the mixture is decreased, the concrete is air-entrained or if the slump flow level is changed, the exact bleed relationship will likely change. However, this gradation information is still useful for quality control and mixture proportioning guidance.

Packing Density

The packing density will impact both the paste requirement and the mortar rheology.

Particle Shape

More-rounded, equidimensional particles will flow more easily than angular and irregularly shaped materials.

Characteristics of Combined Aggregates

Most contemporary concrete technologists will consider the combined characteristics of the aggregates when proportioning mixtures. For SCC, the combined aggregate gradation is used most often for the initial selection of aggregate sources and for developing or adjusting mixture proportions. It can be analyzed from different perspectives and using different tools such as:

- A scatter plot of the individual percent retained on each sieve. In this scenario, the technologist is looking for large gaps between particle sizes. If large gaps exist, the technologist may change the ratio of the aggregate blend or add other aggregates to the blend, depending on availability. Some have proposed the use of the 8–18 rule when analyzing the percent-retained graph. This rule requires each sieve, except for the uppermost screen and then those screens below 300 μm, to retain between 8% and 18% of the aggregate. The other sieves may contain less than 8%. This rule is an attempt to limit large gaps.
- The combined aggregate grading can be compared against a numerical ideal curve such as the power 0.45 curve, Bolomey's curve, or others.[25] Then adjustments to the aggregates used or the ratio can be made to more closely fit the ideal.

In addition to the combined gradation, researchers have suggested measuring the void content of the combined aggregates.[26,27] The combined void content is then used in the proportioning process to establish the required paste content of the SCC mixture. The typical goal is to determine the aggregate combination with the lowest void content so as to reduce the required paste content.

The concept of looking at the properties of combined materials, in particular the particle size distribution, can be extended to include all solid particles in the mixture, including powders. Although it is not within the ability of most concrete producers to quantify the size distribution of powders, knowing in general how the materials relate to one another is conceptually useful. The average particle size decreases from coarse to fine aggregates; then Portland cement, fly ash, and slag cement are similar in size but smaller than the aggregates; and finally there is silica fume, which has the smallest average particle size. The fine and coarse aggregates overlap, and the primary powders are close to the bottom end of the fine aggregate distribution. There exists a particle size gap between silica fume and the other powders. It is here that materials such as milled and ground limestone can be processed to fit specific gaps within the particle size range from silica fume to fine aggregate. Fitting the solid materials together appropriately will have a significant impact on both the water and HRWR demand and the fresh properties of an SCC mixture. Because the distribution of particle sizes for a material changes based on the material source, theoretically each set of materials will have its own optimum blend. To the author's knowledge, at this point no perfect model exists that can take the physical characteristics of all constituent materials, enter them into a program, and accurately predict performance. The author is aware, however, of current attempts to develop this type of modeling, which may produce fruit in the future.[28]

Summary

The development and control of SCC mixtures starts with understanding the influence of each constituent material. One then considers the interactions of materials as they are combined and/or substituted for one another, asking questions along the way such as "What happens when fly ash replaces cement?" or "What if I replace this fine aggregate with another one?" The overriding principle in this analysis is to compare the materials based on projected influence on mixture performance. Then, through the mixture proportioning and evaluation process, with the chosen materials, one will make the finer adjustments in order to promote the optimum fresh and hardened properties.

Chapter 6

Chemical Admixtures

Introduction

An admixture is a material other than water, aggregates, cementitious materials, and fiber reinforcement, used as an ingredient of a concrete mixture to modify its freshly mixed, setting, or hardened properties and that is added to the batch before or during its mixing.[1] Numerous types of chemical admixtures exist, most of which can be used in SCC. Certain admixtures, however, are used more often than others and these are shown in Table 6.1, along with the general reasons for their use.

When SCC was first introduced into the North American market, there was a misconception that it was a chemical admixture. Some thought this admixture could make SCC out of any set of mixture proportions. This is

Table 6.1 Chemical Admixtures Typically Used in SCC

Admixture Type	Reason for Use
High-range water reducer (HRWR)	Minimize water content to ensure adequate viscosity. Adjust dosage to increase or decrease slump flow.
Viscosity-modifying admixture (VMA)	Enhance viscosity to promote greater mixture stability and reduce bleeding.
Workability-retaining admixture	Provide controlled workability/slump flow retention without retardation.
Accelerating admixture	Increase early age compressive strength development. Facilitate normal setting in cold temperatures.
Retarding and hydration control admixtures	Slow the rate of cement hydration to delay setting in hot temperatures and extend workability time.
Air-entraining admixture (AEA)	Enhance freeze/thaw durability, and increase mixture paste content to promote flow and stability.

certainly not the case. It does, however, highlight the fact that when one thinks of SCC, one also thinks of chemical admixtures. Admixtures are used to control very specific characteristics of the mixture, and combinations of admixtures are used to modify multiple properties simultaneously. For example, one can use a high-range water reducer (HRWR) and a viscosity-modifying admixture (VMA) to increase the slump flow and enhance the stability of an SCC mixture, respectively. This is one of the general characteristics that make chemical admixtures such powerful tools. Unlike increasing the water content to increase flow, theoretically, an HRWR can be added to an SCC mixture to increase fluidity with no other adjustment needing to be made. When water is increased, however, the added volume must be compensated for by decreasing some other component in the mixture, thereby potentially affecting other properties. The same could be said for the use of VMAs versus other proportioning techniques for increasing or decreasing SCC mixture viscosity. Not all chemical admixtures are necessary when producing SCC; however, HRWR admixtures are. Quality SCC cannot be made without the use of this admixture type. Other admixtures, such as VMAs and workability-retaining admixtures, are useful in many circumstances, but may not always be necessary.

High-Range Water Reducers

HRWR admixtures are the most important chemical admixtures used in the production of SCC. Their primary function is to disperse cement particles. By doing so, they provide high levels of fluidity without the use of water, thereby lowering the potential for bleeding and segregation. In addition to controlling the mixture fluidity, this ability to change water content by adjusting HRWR dosage allows the practitioner to manipulate SCC mixture viscosity to suit a given application.

Over the years, different dispersant chemistries have been developed, with various levels of dispersing strength. Some of the first and most widely used HRWRs were based on naphthalene chemistry. These products are still used in many parts of the world. Their use in SCC is somewhat limited, however, primarily because of the dispersant dosage required to achieve SCC fluidity. These higher dosages may result in a delayed rate of hardening and delayed early strength development, which are undesirable properties in many applications. Melamine-based HRWRs suffer from more rapid slump loss, which limits their use because maintaining workability is critical for SCC mixtures.

Polycarboxylate ether (PCE)-based HRWRs are not limited by dosage inefficiency, retardation, or workability retention issues like the other technologies mentioned, and can be applied more generally across segments and applications. The advent of PCE-based dispersants has facilitated the worldwide awareness and development of SCC.

Polycarboxylate Ether (PCE) Technology

Historically, the ability to modify the structure of dispersant molecules was limited, but the introduction of PCE dispersants has changed that. Whereas previous dispersants were mainly by-products of other industries, it is now possible for a variety of molecules to be developed for the sole purpose of dispersing Portland cement.

PCEs are comb polymers, meaning they are characterized as having a backbone with pendant side chains much like the teeth of a comb. Along the molecule's backbone are negatively charged binding sites (Figure 6.1). The structure of the molecule can be manipulated so as to provide various performance characteristics to the concrete mixture. For example, the molecular weight of the backbone and/or side chains can be altered, making them larger or smaller, and/or the concentration of side chains or charged sites along the backbone can be adjusted.

PCE-based HRWRs disperse cement grains by two complementary mechanisms: electrostatic repulsion and steric hindrance. Electrostatic repulsion is the mechanism by which the dispersant molecule adsorbs to the cement grain, with the binding sites along the backbone creating a negatively charged surface. The cement grains are then repelled from one another owing to the like charges on their surfaces (Figure 6.2). This is the mechanism by which most dispersants have historically functioned.

PCEs also disperse Portland cement grains via steric hindrance, which is more of a physical mechanism. The pendant side chains, which act similarly to arms coming off of the cement grains, hinder the re-agglomeration of the cement particles (Figure 6.3).

PCE Structure/Performance Relationship

The ability to control synthesis of PCE structure allows a chemist to design a variety of dispersants, such as one that is an excellent water reducer or one that may maintain high levels of workability over longer periods of time. *It is important to note that not all PCE-based HRWRs are the same, and one must know the characteristics of the product one intends to use.* By changing the structure of the PCE molecule, certain performance characteristics can

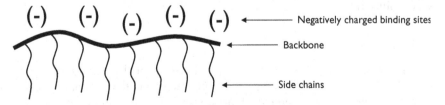

Figure 6.1 Basic Comb Polymer/PCE Structure.

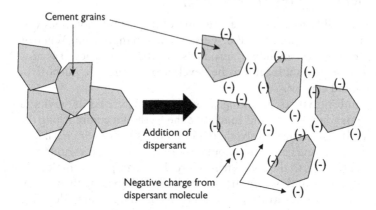

Figure 6.2 Electrostatic Repulsion of Cement Grains by Dispersant Molecules.

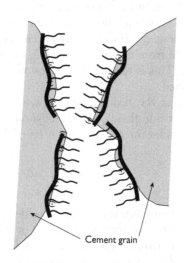

Figure 6.3 Steric Hindrance Dispersing Cement Grains.

be modified, such as dispersion efficiency, concrete mixture viscosity, and workability retention.

DISPERSION EFFICIENCY

This is measured as either the increase in fluidity or the decrease in water per unit change in dispersant dosage. It is a characteristic that ultimately leads to product dosage requirement. A product with a more moderate dosage requirement provides a higher level of control to the producer. This allows smaller dosage adjustments to be made, as needed, during production. It is more difficult to make precise adjustments with very low-dose products. Figure 6.4 shows the dosage responses of three different PCE-based HRWRs.[2] The dosage efficiency of PCE 2 is greater than those of PCEs 1 and 3. A product with a dosage response similar to that of PCE 2 is more easily used in locations where the types of conventional and SCC mixtures being produced are relatively limited, such as in precast plants. The dosage responses of PCEs 1 and 3 are more likely to be used in locations where a broader range of mixture types are produced, such as in ready-mixed concrete plants.

CONCRETE MIXTURE VISCOSITY

The viscosity of an SCC mixture influences several other characteristics, including the passing ability, stability, and surface finish of the concrete. The structure of a PCE molecule can influence the viscosity of an SCC mixture. Table 6.2 shows data comparing the effect on viscosity of three HRWRs based on three different PCE molecules.[3] The rheological parameters were measured using the IBB concrete rheometer. Note the range in viscosity values achieved by changing the HRWR. The viscosity shift is also reflected in the T50 times: PCE 2 has the highest viscosity and T50 time, while PCE 1 has the lowest viscosity and T50 time. The impact of viscosity on segregation,[4] bleeding,[5] and VSI[6] is also presented. For a given mixture at a constant slump flow value, as the viscosity increases, the mixture stability increases.

WORKABILITY RETENTION

This is critical for SCC mixtures, and without it the benefits associated with SCC can be lost. Workability retention can be influenced by the molecular

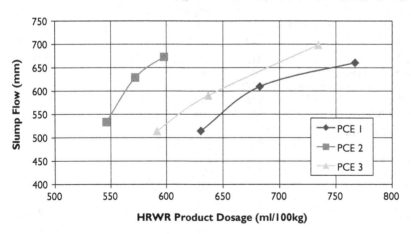

Figure 6.4 Dosage Responses of Three PCE-Based HRWRs.

structure of the polymer used in a given PCE-based HRWR. Table 6.3 compares two PCE-based HRWRs and their workability retention over 80 minutes.[7] Note the tendency for PCE 3 to begin losing workability within the first 30 minutes, while PCE 4 does not begin losing workability until after 55 minutes. It should be noted, however, that this workability retention (or loss) profile is influenced by the mixture proportions, the composition of the powder fraction of the SCC mixture, and the presence of other admixtures, among other things.

Table 6.2 PCE Influence on Viscosity and Stability

HRWR	PCE 1	PCE 2	PCE 3
Dosage (ml/100 kg)	780	559	676
Cement (kg/m³)	434	429	429
Coarse Aggregate (kg/m³)	973	961	961
Fine Aggregate (kg/m³)	865	853	853
Water (kg/m³)	180	177	177
Air (%)	1.3	1.3	1.5
Yield (g)	0.200	0.100	0.100
Viscosity (h)	3.600	5.400	4.600
Slump Flow (mm)	660	660	660
Segregation Factor (%)	11.7	2.8	10.6
Bleed (%)	1.28	0.06	0
T50 (s)	1.8	3.5	2.2
VSI	2	1	1.5

Table 6.3 Workability Retention Comparison Between Two Different PCE-Based HRWRs

HRWR	PCE 3	PCE 4
Dosage	520	585
Type I Cement (kg/m³)	340	339
Class F Fly Ash (kg/m³)	60	60
Coarse Aggregate (kg/m³)	1076	1070
Fine Aggregate (kg/m³)	827	822
Water (kg/m³)	160	160
Slump Flow (mm)		
Initial	675	660
30 Minutes	535	660
55 Minutes	420	660
80 Minutes	335	580
Air (%)		
Initial	0.9	1.4
80 minutes	1.7	1.6

PCEs also impact the rate of hardening and early compressive strength development of SCC mixtures differently. Some promote more rapid setting and rapid strength gain while others promote normal strength gain. Early strength is important for precast/prestressed concrete producers as it determines how quickly de-molding can occur. Rapid setting or rapid strength gain is also useful as one incorporates more supplementary cementitious material into the SCC mixture proportions.

Interaction with Portland Cement

HRWR admixtures function by adsorbing to and dispersing Portland cement grains. The interaction of PCE dispersants and the cement grain surface is highly complex. Suffice it to say that if one randomly changes the cement, a resulting change in initial SCC fluidity and/or HRWR dosage requirement may occur. The degree of change can range from minimal to significant. Figure 6.5 shows the range in the required dosage of a single PCE HRWR to achieve a similar workability in over 40 different Portland cements available in North America.[8]

One should follow the recommendations of the admixture supplier regarding the use of a particular HRWR. Discussing the performance characteristics of HRWRs with one's chemical admixture representative is recommended in order to understand the relative performance differences between the available products.

Viscosity-Modifying Admixtures

SCC evolved from underwater concreting technology.[9] When concrete is placed underwater, it must be highly workable, and resistant to segregation and the washing out of cementitious fines. Anti-washout admixtures were developed to increase the cohesiveness and viscosity of underwater concrete

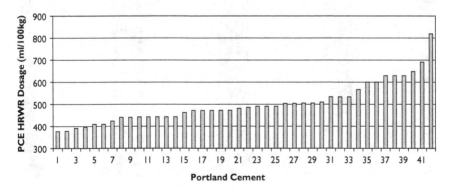

Figure 6.5 Portland Cement Influence on PCE HRWR Dosage.

mixtures and eliminate or significantly decrease the separation of mixture constituents during placement. The admixtures became the first of what are now known as viscosity-modifying admixtures (VMAs).

VMAs are used to enhance stability and provide robustness to SCC mixtures. There are a variety of VMAs: some thicken the cement paste, while others specifically thicken the water. Figure 6.6 shows the water-thickening effect of one particular VMA. This photograph shows four cylinders, 90 mm in diameter and 460 mm in height. The leftmost cylinder is filled with water, while the other three are filled with different water/VMA mixtures, resulting in highly fluid mixtures with progressively higher viscosities. The photograph was taken approximately 25 seconds after a golf ball was dropped into each cylinder. Note the varied settlement depths of the golf balls. As more VMA is added, the mixture viscosity is increased and settlement decreases.

The effect on mixture viscosity, bleeding, and aggregate segregation is dependent on the VMA chosen. Some are more potent bleed control products, while others will more effectively reduce coarse aggregate segregation. Table 6.4 shows the impact of three different VMAs on mixture stability.[10]

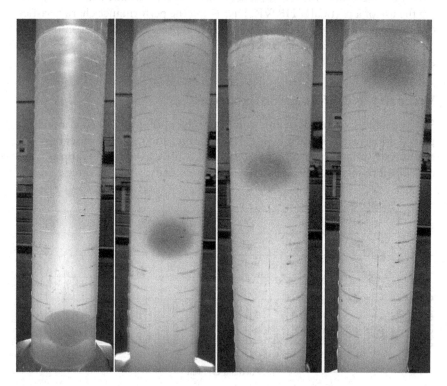

Figure 6.6 Golf Ball Settling in Solutions of Different Viscosity.

Table 6.4 VMA Impact on Stability

	Reference	VMA 1	VMA 2	VMA 3
Cement (kg/m³)	422	426	430	426
Coarse Aggregate (kg/m³)	945	954	964	954
Fine Aggregate (kg/m³)	840	848	857	848
Water (kg/m³)	200	203	205	203
w/c Ratio	0.48	0.48	0.48	0.48
HRWR (ml/100 kg)	1021	1268	1066	1066
VMA (ml/100 kg)	0	260	650	650
Slump Flow (mm)	826	787	730	781
T50 (s)	0.9	1.3	1	0.7
Bleed (%)	13.3	0.4	2.3	10.5
Segregation Factor (%)	49	37	24	36

The SCC mixture used in this evaluation was purposefully developed to be unstable. In this way, the impact of each VMA is more clearly demonstrated. To purposefully promote instability, a relatively higher water/cement (w/c) ratio, lower sand and cement content, and higher slump flow value were incorporated. All of the mixtures in this evaluation utilized the same standard ASTM Type I cement, fine aggregate, coarse aggregate, and PCE-based HRWR. The VMAs were dosed according to the manufacturer's recommendation. Segregation was measured using the column test[4] and bleeding was measured using ASTM C 232.[5]

Note the highly unstable reference mixture: both the bleed percentage and the column segregation percentage are very high, as intended. As the different VMAs are incorporated into the SCC mixture, the segregation factor decreases. Some of this can be attributed to a slight decrease in slump flow; however, the primary driver is the VMA chemistry. The relative reduction in aggregate segregation versus the reference mixture varies by product. Figure 6.7 presents the percentage reduction in segregation relative to the reference mixture by VMA.

The three VMAs also provide different degrees of bleed reduction. Figure 6.8 presents the percent reduction in bleed relative to the reference mixture.

The various VMAs have certain strengths and weaknesses. They are tools to be used during SCC mixture development and their selection depends on the characteristics of the available materials. Because VMA 1 is highly effective in reducing bleed, and reasonably effective in reducing aggregate settlement, it would be useful in proportioning mixtures incorporating coarser and angular fine aggregates, such as those with limited contents of material passing the 300 µm sieve and in mixtures where one prefers to keep the powder content low. VMA 3 provides a moderate reduction in both bleed and aggregate segregation and is useful for providing stability to mixtures

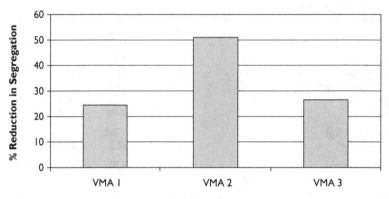

Figure 6.7 Percent Reduction in Segregation Relative to Reference Mixture.

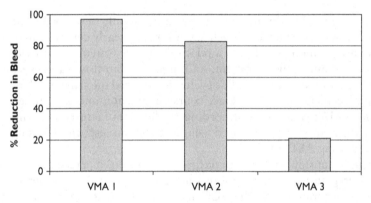

Figure 6.8 Percent Reduction in Bleed Relative to Reference Mixture.

that are in need of a slight to moderate stability improvement. VMA 2 significantly reduces both bleeding and aggregate segregation; it is a more robust product and would be useful in a wide range of SCC mixtures.

VMAs and Robustness

Robustness is defined as the degree of insensitivity of an SCC mixture during production to fluctuations in raw material characteristics such as moisture content, particle size distribution changes, and other things that can lead to changes in the mixture's fresh properties. Robustness can be assessed by measuring changes to any of the fresh properties of SCC, including slump flow, T50 time, segregation, bleeding, passing ability, or others. VMAs have been shown to be useful in providing robustness to the slump flow and passing ability of SCC mixtures by a number of different researchers.[11,12]

The following experiment evaluated the impact of VMAs on SCC robustness.[10] The experiment was developed to simulate the condition of free moisture fluctuation in fine aggregate. A reference mixture and mixtures containing three different VMAs were tested. All of the mixtures in this evaluation utilized the same standard ASTM Type I cement, fine aggregate, coarse aggregate, and a PCE-based HRWR. The VMAs were dosed according to the manufacturer's recommendation and the mixtures were tested for slump flow, VSI, T50 time, bleed percentage, and segregation percentage. Each mixture was run to an initial slump flow of 625–675 mm and then re-run three times, with the only adjustment being the addition of 8, 16, and 24 kg/m^3 of water to each mixture. These water additions correspond to approximately 1%, 2%, and 3% of the fine aggregate mass. In this way, the water addition could be likened to a fine aggregate moisture fluctuation. Table 6.5 shows the mixture proportions and fresh properties of the initial mixtures.

Figures 6.9–6.12 show the robustness of all four mixtures in terms of slump flow, T50 time, bleed percentage, and segregation percentage as a function of the percent increase in fine aggregate moisture.

The reference, non-VMA mixture has the greatest ultimate change in slump flow. The three VMAs all provide robustness to the slump flow, but to differing degrees, with VMA 2 being the most effective across all water additions.

It is the author's experience that the precision of the T50 test decreases as the time decreases and this is particularly true when the time falls below 1 second. At these very short times, the ability to start and stop the stopwatch varies between operators and the ability to effectively discriminate between mixtures drops considerably. Notice the initial change in T50 time with the first and second additions of water. It is at these points that we clearly see the robustness provided by the VMAs. In this series, VMAs 1 and 2 seem to provide the greatest ability to maintain viscosity as additional water is added, although all three showed improvement over the reference.

Table 6.5 Initial Mixture Proportions and Fresh Properties for Robustness Evaluation

	Reference	VMA 1	VMA 2	VMA 3
Cement (kg/m^3)	432	436	436	434
Coarse Aggregate (kg/m^3)	969	978	978	974
Fine Aggregate (kg/m^3)	861	869	869	865
Water (kg/m^3)	178	181	181	180
w/c Ratio	0.41	0.41	0.41	0.41
HRWR (ml/100 kg)	1021	1268	1066	1066
VMA (ml/100 kg)	0	260	650	650
Slump Flow (mm)	641	648	667	641
VSI	1	0	1	0
T50 (s)	2.6	3	2	2.4
Bleed (%)	0.4	0	1.1	0.6
Segregation Factor (%)	22	7	16	18

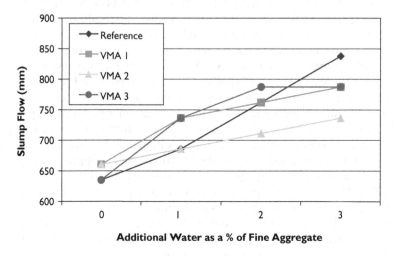

Figure 6.9 Slump Flow Robustness.

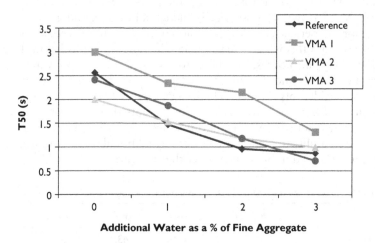

Figure 6.10 T50 Robustness.

The influence of the various VMAs on controlling bleed has already been discussed; however, Figure 6.9 presents further information that supports the previous data. Not all VMAs are the same with respect to how they interact with water. VMAs 1 and 2 provide exceptional bleed robustness as water continues to be added to the system. VMA 3 is not as effective in providing bleed robustness as the other two.

Robustness to aggregate segregation is provided by the use of VMAs, although, as with the other robustness properties measured, the robustness

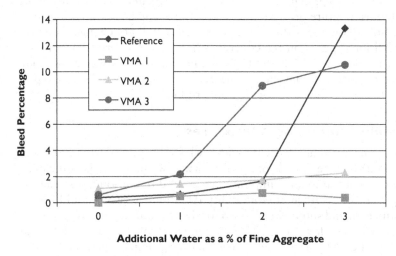

Figure 6.11 Bleed Robustness.

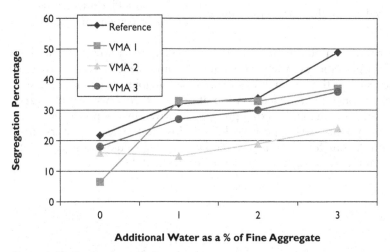

Figure 6.12 Segregation Robustness.

provided varies by VMA type. In this series, we see that VMA 1 initially provides less robustness to the system after the initial increase in water; however, after this initial increase, the segregation percentage stabilizes. This initial increase could be due to the presumed lower yield stress of the mortar fraction as a result of the increase in HRWR over the reference. VMA 2 in this case provides very good and consistent robustness to the SCC mixture tested.

The relationships presented as a result of this experiment will change if the raw materials, mixture proportions, or starting slump flow levels change. It should also be noted that the inclusion of VMAs will not change the effect of the additional water on the concrete's hardened properties. The point, however, is that, like PCE-based HRWRs, not all VMAs are created equal, and it is recommended that the producer consult with the admixture representative to choose the most appropriate product.

Workability-Retaining Admixtures

All mixtures have a window of time within which the original fresh properties are maintained. Once outside of this time window, an SCC mixture will be reduced to a conventional slump mixture in need of consolidation energy. The importance of sufficient workability retention for SCC should not be underestimated, and should be planned for. For example, a precast producer may need less time for transport and completion of placement than that required for a cast-in-place project. The following simple equation can help one determine the amount of workability retention needed:

$$\text{Required slump flow retention time} = \frac{\text{Transit time + Placement time +}}{\text{Time for placement of next batch}}$$

Transit time and placement time are self-explanatory; however, the time required for placement of the next batch is the time needed to ensure that no pour lines, such as those seen in Figure 6.13, or cold joints occur as successive deliveries of concrete are placed on top of one another. This time can vary, based on the batching and mixing process employed by the producer.

Concrete practitioners expect workability loss and, when necessary, prepare to compensate for it in one of several ways developed over the years;[13,14] not all techniques, however, are suitable for use with SCC:

Figure 6.13 Pour Lines as a Result of Inadequate Workability Retention.

- Re-tempering with water at the jobsite. In the author's experience, this method is rarely, if ever, used with SCC, nor should it be. With SCC, additions of water should be limited to adjustments at the batch plant by a single individual responsible for batching and quality. Water has a direct and significant impact on the mixture viscosity, bleeding potential, and segregation potential, and should be tightly controlled.
- Re-dosing with HRWR at the jobsite. This method is occasionally used in practice, and a dedicated person, responsible for adding the HRWR at the casting site, is required. If one person is not assigned this responsibility, the consistency and quality of the SCC deliveries may be compromised.
- Batching to a higher-than-target slump flow to compensate for slump loss during transit. In the author's experience, in air-entrained SCC, this method is not used very often, as it may require one to batch a mixture to a fluidity level that could negatively impact the initial stability and air content.
- The addition of a hydration-controlling or -retarding admixture to slow hydration and stiffening. The method of using a retarding or hydration control admixture to extend workability is a common method used in practice. Most concrete producers have some experience and history using these types of admixtures.

Workability-retaining admixtures are a new class of admixture designed to only influence workability retention time. Figure 6.14 presents the slump flow retention of a reference mixture versus three mixtures containing increasing dosages of a workability-retaining admixture.[15] Each mixture had an initial slump flow of 600–675 mm.

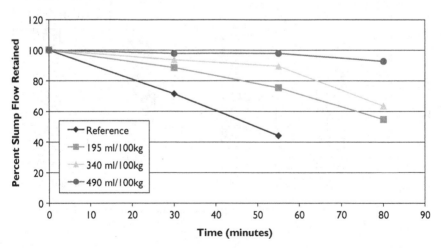

Figure 6.14 Workability Retention Flexibility.

The flexible dosage allows the concrete producer to respond to changes in temperature, materials, and project conditions and to achieve consistent mixture performance. The performance is achieved without slowing cement hydration, as shown in Figure 6.15, which presents the rate of hardening for the same four mixtures.

Air-Entraining and Other Admixtures

Air entrainment in SCC is possible and is a daily occurrence in multiple locations across North America. The dosage of air-entraining admixture needed for a specified percentage of air will vary. Similar to conventional concrete, it is influenced by a number of factors, including mixture proportions, fine aggregate grading, mixer type and mixing efficiency, and the presence of other admixtures, including HRWRs. Particular to SCC is the influence of mixture stability. As a mixture becomes less stable, it becomes more difficult to consistently entrain air. The concrete producer should evaluate an SCC mixture to ensure its stability when air entrainment is required. At the same time, since air increases the paste volume of a mixture, its inclusion can improve the stability of SCC mixtures.

Other admixtures that can and are frequently used in the production of SCC mixtures include normal and mid-range water reducers, corrosion-inhibiting admixtures, liquid color admixtures and others. Most admixtures that are used in the production of conventional concrete can be used in SCC. Always follow the manufacturer's recommendations for combining multiple admixtures in a single concrete mixture.

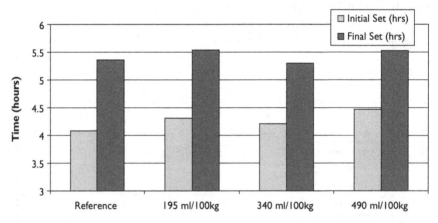

Figure 6.15 Rate of Hardening of Mixtures Containing Workability-Retaining Admixture.

Summary

Chemical admixtures are necessary for the production of good quality SCC mixtures. HRWR admixtures are essential, while other admixtures, such as VMAs and workability-retaining admixtures, are used regularly. Not all PCE-based HRWRs are the same. They may be different in their dosage efficiency, mixture viscosity effect, workability retention, and influence on rate of hardening and early strength development. Likewise, not all VMAs are the same. They can differ in their influence on bleeding, segregation, and robustness. It is very important that the concrete producer and contractor clearly determine the necessary performance attributes of the SCC mixture. The concrete producer should then consult the admixture representative to ensure that the admixtures in use are the most appropriate for the application and the materials available.

Chapter 7

Considerations Prior to Mixture Proportioning

Introduction

Similar to conventional concrete, the successful use of SCC requires consideration of the project requirements and placement techniques prior to the selection of fresh property targets and mixture proportions. For those lacking experience with SCC, the process may seem complex. It only requires, however, an understanding of those project attributes that are most important and how to design the SCC mixture accordingly. This understanding requires setting aside some preparation time before the beginning of a project. During preparation, issues and subsequent adjustments will be identified, thereby minimizing time-consuming trial-and-error testing. The preparation process considers a number of factors, including understanding the required hardened properties and knowing the characteristics and costs of available raw materials. Hardened property requirements may impose limits on the types and quantities of materials used. For example, if high early compressive strength is required, the water/cement (w/c) ratio will be set at a relatively low level, thereby increasing the mixture viscosity. In this case, the ability to adjust mixture proportions to influence mixture viscosity is limited owing to the compressive strength requirement. Raw materials and their influences were discussed in Chapter 5. The impact of specific materials, for example coarse aggregate particle shape, needs to be determined. More-angular aggregates impede flow and typically have a lower packing density than rounded aggregates, which in turn influences the paste requirement and mixture cost. If a round aggregate is available, but is from a source further away, are the benefits provided by the rounded aggregate worth the additional freight cost and the sustainability impact of the increased fuel usage?

The primary activity in preparing for proportioning an SCC mixture, however, is to determine the fresh SCC performance targets for the project. If a performance history or other experience with which to establish the performance targets is not available, how does one decide on the level of slump flow or the importance of passing ability? This chapter presents a fresh property analysis of over 65 case studies from the literature. Following this analysis, industry guidelines from Japan, Europe, and North America are presented in some detail.

A Review of Case Studies

In a previously published case study analysis, a wide range of fresh property values were reported. A slump flow range of 500 to over 800 mm and highly variable flow rate measurements, such as the T50 and V-funnel tests, were reported.[1] This trend of a wide slump flow and plastic viscosity range for worldwide SCC mixtures has also been confirmed by Wallevik.[2] This reaffirms what has previously been said, namely, that SCC mixtures can have a wide range of performance attributes. For the current analysis, the author collected information on over 67 SCC applications published between 1999 and 2010, including a number of those referenced in an earlier analysis.[1] The projects occurred in 16 countries, with 42 cases being for ready-mix (cast-in-place) applications, 19 for precast applications, and 6 not classified. This sample of 67 case studies over an 11-year period is, by all intents, a small sample of projects. The value of this analysis, therefore, lies not in establishing "hard" specifiable numbers, but rather in understanding the relative change in fresh SCC properties as influenced by project requirements. Table 7.1 presents the fresh property data from these case studies.

Slump Flow Analysis

All projects reported a slump flow value. The average slump flow was 670 mm and the range was 525–840 mm. But what project conditions require the use of a 525 mm versus an 840 mm slump flow? The answer to this question has a direct impact on mixture cost because, all other things being equal, as slump flow performance increases, mixtures become more expensive because high-range water-reducer (HRWR) dosage increases, the paste and therefore powder content requirement is higher, a viscosity-modifying admixture (VMA) may be necessary, and greater quality control resources for material control may be necessary. In some ways, slump flow may be thought of as analogous to compressive strength, meaning that a wide range of values are possible with both properties. In each case, one could choose to use a higher-performing mixture to satisfy the requirements of a lower performance specification. For example, one could use a 40 MPa mixture on a project only requiring 20 MPa strength. Likewise, one could choose to use a 750 mm slump flow mixture when in reality just a 600 mm slump flow is necessary. These, however, would not be cost-effective approaches. One should choose the level of performance needed and develop a mixture to achieve that target.

The distribution of slump flow by segment (ready-mix or precast) provides some insight into the application of slump flows. Figure 7.1 shows the

Table 7.1 Slump Flow Case Studies

Ref.	Location	Segment	Faster Speed of Construction and Time Savings	Labor Savings	Ease of Placement and Consolidation Through Dense Reinforcement	Ease of Placement and Consolidation in a Complex Structure or Shape	Ease of Placement with Limited Access	Provide Higher In-Place Quality and Aesthetics	Improved Worker Safety and Noise Reduction	Slump Flow (mm)	T50 (s)	V-Funnel(s)	O-Funnel(s)	U-Box	L-box, h2/h1
3	Japan	CIP				X	X			650					
3	Japan	CIP					X			650					
3	Japan	CIP	X		X				X	700					
3	Japan	CIP	X	X						650					
4	Japan	CIP	X							650	5		10	R3	
5	Sweden	CIP	X					X	X	650					
6	Japan	CIP						X	X	600					
7	Austria	CIP					X			725					
8	USA	CIP					X			710					
8	USA	CIP					X	X		610					
8	USA	Precast		X	X	X		X		560					
8	USA	CIP					X			660					
8	Canada	CIP								640		3.8			
9	Japan	CIP			X		X			650	6			R1	
10	Japan	CIP			X		X			717	4.8			R1	
10	Japan	CIP					X			667	5.9			R2	
11	Korea	CIP					X			650	5	15		R2	
12	Norway	CIP							X	715					
13	Netherlands	CIP				X		X		730		8.5			
13	Netherlands	CIP			X	X	X			700		12			
14	Canada	CIP								660					
15	Argentina	CIP					X	X		680	1.8			Yes	0.96
15	Argentina	CIP			X		X			660	2			Yes	0.93
15	Argentina	CIP					X			700	2.6			Yes	0.9
16	USA	Precast		X	X		X	X	X	650					
17	Taiwan	CIP			X					650					
18	USA	Precast	X	X				X	X	525					
19	S Africa	CIP					X	X	X	650					

#	Country	Type	1	2	3	4	5	Value	a	b	c
19	S Africa	CIP	X			X		600	1	6	0.8
19	S Africa	CIP				X		750			
20	Portugal	CIP	X			X		650	2.6	14	0.87
20	Portugal	CIP						750	2.1		
21	New Zealand	CIP		X			X	700			
22	Italy	CIP		X			X	675			
22	Italy	CIP		X			X	730			
23	Canada	CIP			X		X	650		3.95	
24	Chile	CIP			X		X	720			
25	Sweden	CIP			X		X	750			
26	Italy	CIP		X	X	X	X	600	8	8	
27	Canada	CIP						700			
27	Canada	CIP						695			
27	Canada	CIP						700			
27	Canada	CIP						640			
28	USA	CIP		X	X	X	X	650			
29	USA	CIP				X	X	650			
30	USA	Precast						660			
30	USA	Precast						710			
30	USA	Precast						560			
30	USA	Precast						660			
30	USA	Precast						585			
30	USA	Precast						610			
30	USA	Precast						710			
30	USA	Precast						660			
8	USA						X	660			
8	USA							660			
8	USA							660			
8	USA							660			
8	USA							840			
8	USA							685			
31	Mexico	Precast	X	X			X	685		15	1
31	Mexico	Precast					X	650			
31	Mexico	Precast	X				X	705			
31	Mexico	Precast					X	655			
31	Mexico	Precast						750			
31	Mexico	Precast	X	X				650			
31	Mexico	Precast	X	X				600			
31	Mexico	Precast						700			—

number of occurrences for discreet slump flow ranges by segment and Figure 7.2 shows these occurrences as a percentage by segment. According to these data, the ready-mixed segment prefers to use mixtures with slump flow higher than 600 mm. Although both segments primarily used mixtures with slump flow in the 650–750 mm range, precast concrete producers were able to consistently use mixtures with lower slump flows: slump flows less than 650 mm were used 30% of the time and slump flows less than 600 mm 20% of the time. The reason for this is not immediately evident; however, SCC acceptance is at a more advanced state in the precast industry.[32] It may be that precast producers have learned and established the appropriate performance levels for their range of applications. Although it is not part of this analysis, it should also be noted that in certain countries, Denmark in particular, the slump flow target for ready-mix delivered, cast-in-place, horizontal applications is regularly 550 mm. In 2007, 30% of ready-mix concrete production in Denmark was SCC.[33] In other words, where SCC has a high level of acceptance, and those using it have more experience, whether by segment or geography, mixtures with lower slump flow values are used consistently. It should be noted that in order to use a lower-slump-flow SCC, one must ensure that the viscosity is also kept relatively low, to ensure filling ability.

The slump flow data were further analyzed to investigate the correlation between the slump flow level and the benefits received from using SCC. The benefits of using SCC have been previously presented and include the following:

- Faster speed of construction and time savings
- Labor savings
- Ease of placement and consolidation through dense reinforcement

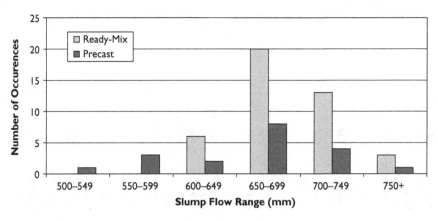

Figure 7.1 Slump Flow Occurrences by Segment.

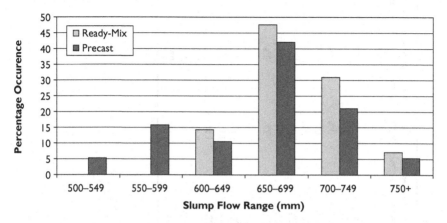

Figure 7.2 Percentage of Time a Slump Flow Range was Used by Segment.

- Ease of placement and consolidation in a complex structure or shape
- Ease of placement with limited access
- Higher in-place quality and aesthetics
- Improved worker safety and noise reduction.

Table 7.2 provides data showing the percentage of time a slump flow range was used to achieve a specific benefit. Separate data are presented for ready-mix and precast applications. The ready-mixed data show a potentially interesting trend. Figure 7.3 shows the distribution of slump flow by performance benefit for ready-mix applications. Faster speed of construction and time savings, as well as labor savings, are clear economics-driven benefits; although most benefits will be reduced to a financial benefit at some point, the others can be classified as performance-driven or technical benefits. What Figure 7.3 suggests is that when SCC is used for purely economic reasons such as for reducing labor, the slump flow value is not required to be as high as when a technical or performance need is present. This seems logical, because the number of laborers required for consolidation decreases significantly as concrete performance shifts from conventional slump to SCC; however, once at the SCC performance level, the required number of consolidation laborers, if any, cannot be reduced further regardless of how much the slump flow is increased. Similarly, the time savings when using SCC are primarily due to the relatively continuous nature of the placement without pausing for consolidation. With consolidation effort eliminated, no real additional time benefit (for placement and consolidation) can be obtained by using a higher slump flow. In contrast, if the project involves a very complex structure or shape, Figure 7.3 suggests that the trend is toward higher slump flow values in the 700–749 mm range.

Table 7.2 Slump Flow Level by Benefits Achieved

	Faster Speed of Construction and Time Savings	Labor Savings	Ease of Placement and Consolidation Through Dense Reinforcement	Ease of Placement and Consolidation in a Complex Structure or Shape	Ease of Placement with Limited Access	Provide Higher In-Place Quality and Aesthetics	Improved Worker Safety and Noise Reduction
Ready-Mixed							
500–549 mm	0%	0%	0%	0%	0%	0%	0%
550–599 mm	0%	0%	0%	0%	0%	0%	0%
600–649 mm	40%	33%	17%	13%	12%	7%	25%
650–699 mm	20%	33%	58%	38%	52%	50%	25%
700–749 mm	20%	33%	25%	50%	28%	36%	50%
≥750 mm	20%	0%	0%	0%	8%	7%	0%
Precast							
500–549 mm	20%	14%	0%	0%	0%	11%	50%
550–599 mm	0%	14%	0%	0%	0%	11%	0%
600–649 mm	0%	14%	50%	0%	0%	11%	0%
650–699 mm	60%	43%	0%	0%	0%	44%	50%
700–749 mm	0%	14%	50%	0%	0%	11%	0%
≥750 mm	20%	0%	0%	0%	0%	11%	0%

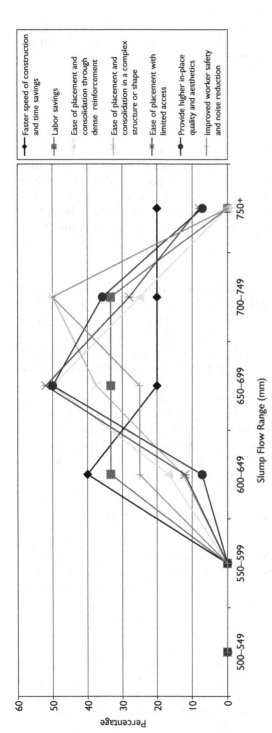

Figure 7.3 Distribution of Slump Flow Values by Performance Benefit.

The precast data, on the other hand, do not show these trends. Both economic and technical/performance benefits are achieved across all of the slump flow ranges.

Viscosity Analysis

A variety of tests and results linked to mixture viscosity were reported. Eleven studies reported T50 time values, with an average of 3.5 s and a range of 1–6 s. Nine studies reported a V-funnel time, with an average of 9.5 s and a range of 3.8–15 s. These ranges show the broad spectrum of viscosity in SCC mixtures. No trend was found to relate a particular viscosity to any SCC benefit. As mentioned previously, the viscosity of a mixture can be dictated by hardened property requirements rather than fresh property requirements.

Passing Ability Analysis

Eight references were found for the U-box test: two used Rank 1, two used Rank 2, one used Rank 3, and three did not list the obstacle rank used. Of the eight studies, three used SCC for ease of placement and consolidation through dense reinforcement, and the Rank 1 obstacle was used for two of these projects; an unknown obstacle was used for the third. Six cases used the L-box test and all had $h2/h1$ ratio greater than 0.8. No projects reported using the J-ring.

This case study review clearly demonstrates the wide range of fresh properties possible in real-world SCC mixtures. Considering the precast industry worldwide and the ready-mix industry in Denmark, it seems that the ability to use the full spectrum of slump flows leads to greater use of SCC technology. The question still remains, however: when are lower or higher slump flows, or other performance attributes, necessary?

A Review of Guidelines for Establishing Target Performance

Guidelines for establishing SCC performance targets exist from Japan, Europe, and North America. Some were intended to guide the writing of SCC specifications, while others were created with the intent of educating and guiding the practitioner in the selection of target performance.

Japanese Guidelines

The development of SCC in Japan occurred as a response to the deterioration of reinforced concrete structures. The deterioration problem was reportedly due to a reduction in the number of skilled concrete workers and poor consolidation practices.[34] Several associations in Japan published guidelines in the late 1990s, including the following:

- The Architectural Institute of Japan (AIJ)—Recommended Practice for High-Fluidity Concrete for Building Construction, 1997. In this document, a range of target slump flows are listed as 550, 600, or 650 mm and the slump flow used depends on where in the structure the concrete is to be placed.[35] The document also states that the slump flow shall not be less than 500 mm or greater than 700 mm.
- The Japan Society of Civil Engineers (JSCE)—Recommendations for Construction Practice of High-Fluidity Concrete, 1998. The JSCE provides an example of distinguishing SCC performance based on passing ability using the U-box test. Three ranks of "self-compactability" are established based on the obstacle through which the concrete is required to flow. Rank 1 has the most restrictive obstacle and is used for applications with clear spacing of 35–60 mm and a steel content of 350 kg/m[3]. Rank 2 has a slightly less restrictive obstacle and is used for applications with a clear spacing of 60–200 mm and a steel content of 100–350 kg/m[3]. Rank 3 uses no obstacle and is used for applications with a clear spacing of >200 mm and a steel content of <100 kg/m[3]. General reinforced concrete structures or members should use Rank 2.[36]

European Guidelines

The *European Guidelines for Self-Compacting Concrete*[37] were published in May 2005. Performance classes for flowability based on slump flow, viscosity based on the T50 or V-funnel tests, passing ability based on the number of bars in the L-box, and segregation resistance based on the sieve segregation test are provided for practical guidance.[37] The classes developed are presented in Tables 7.3–7.6.

In addition to the performance classes, guidance for choosing performance based on general applications is provided. Table 7.7 shows the distribution of performance by application. These guidelines segment SCC performance in a very systematic way and allow for the creation of a performance-based project specification.

Table 7.3 European Slump Flow Classes

Class	Slump Flow (mm)	Application
SF 1	550–650	Minimally reinforced open structures (e.g. housing slabs); sections requiring minimal flow distance (piles, deep foundations).
SF 2	660–750	Suitable for many normal applications.
SF 3	760–850	Vertical applications; highly congested reinforcement and complex shapes; need for greater surface finish. Segregation resistance may be more difficult to control.

Table 7.4 European Viscosity Classes

Class	T50 (s)	V-Funnel Time (s)	Application
Class 1	≤ 2	≤ 8	Good filling ability through congested reinforcement; better surface finish but potential for bleed.
Class 2	> 2	9 to 25	May show thixotropic properties and good segregation resistance. Surface finish may be negatively affected.

Table 7.5 European Passing Ability Classes

Class	Passing Ability	Application
Class 1	h2/h1 ≥ 0.80 with two rebars	Structures with a gap of 80–100 mm
Class 2	h2/h1 ≥ 0.80 with three rebars	Structures with a gap of 60–80 mm

Table 7.6 European Segregation Resistance Classes

Class	Segregation Resistance (%)	Application
Class 1	≤20	Thin slabs; flow distance shorter than 5 m; applications with a confinement gap >80 mm
Class 2	≤15	Vertical applications with flow distance >5 m

Table 7.7 Suggested Ranges of Plastic Property Values for Various Applications Based on European Experience

Application	Slump Flow Value (mm)	V-Funnel Time (s)	Passing Ability	Segregation Resistance
Ramps	470–570	9–25	Specify for SF 1 and 2	
Floors and slabs	470–715	3–5		
Walls and piles	540–640	5–9		Specify for SF 3
Tall and slender elements	600–800	9–25		Specify for SF 3

North American Guidelines

The set of guidelines presented in Table 7.8 was developed by the author and published in 1999 in an internal development report on SCC.[38] They were based on early application experience, and presented on numerous occasions throughout the United States and Canada in 1999 and 2000.

Table 7.8 Suggested Ranges of Plastic Property Values for Various Applications

Application	Slump Flow Value (mm)	U-Box Value*	V-Funnel Value† (s)
Slab	460–710	NA	2–7
Architectural section	610–710	NA	3–10
Wall minimum reinforcement	460–660	NA	5–11
Structural column or wall			
densely reinforced	610–710	>235 mm rising height	8–15

*Box size varies from JSCE. †75 mm × 75 mm opening.

These guidelines were intended for the education of the concrete practitioner and to demonstrate the link between fresh properties and application. To simplify the demonstration of this concept, the type of application, such as slab and wall, was used; however, this oversimplified the process and practitioners continued to require further guidance in using SCC. Through experience, it became clear that certain project variables determine the required fresh SCC properties. With this in mind, a new set of guidelines were developed in 2001.[39] Rather than generalize performance by element type, such as slab, column or wall, the authors desired to clarify more fully the element and project attributes. The modified guidelines described specific element attributes that could impact the necessary SCC properties and suggested a relative importance ranking system of low, medium, and high for each attribute. This made the guidelines more flexible and permitted their use for most projects. The fresh properties were also categorized into low, medium, or high performance levels, and a matrix of project attributes and fresh properties was created. These new guidelines were further modified and incorporated into industry documents such as the Interim Guidelines for the Use of Self-Consolidating Concrete in Precast/Prestressed Concrete Institute Member Plants[40] and the ACI 237R-07 Emerging Technology Series Report on Self-Consolidating Concrete.[8] This section details the various project attributes and provides general guidance for selecting fluidity, passing ability, and viscosity targets. SCC mixtures are always intended to be stable; therefore, segregation resistance is not specifically highlighted, although it is referred to in relation to certain project attributes. The attributes highlighted include reinforcement level, element shape and intricacy, element depth, element length (flow distance), importance of the final surface finish, wall thickness, coarse aggregate content, and placement energy. The use of these tables has been helpful to practitioners in choosing initial performance where no previous experience exists and in programs designed for training employees on the practical use of SCC.

Reinforcement Level

This relates to the minimum clear spacing between sections of reinforcement as well as between the reinforcement and the form walls. The reinforcement level will determine the required level of fluidity and passing ability. It will also influence the selection of the maximum size aggregate. Figures 7.4, 7.5, and 7.6 provide examples of low, medium, and high reinforcement levels.

Table 7.9 shows the relationship between the reinforcement level rankings and suggestions for initial fresh property targets. The unchecked boxes represent the suggested performance space. For example, if the reinforcement level is high, the table suggests a high fluidity, a high passing ability, and a medium to high viscosity. Such tables can also be used as guidance for what not to do; again using the high reinforcement level example, a mixture with low fluidity, poor passing ability, and too low of a viscosity should not be used.

Element Shape Intricacy

This rates the complexity of the structure or form shape, giving an indication of the difficulty a mixture will have in filling a form without leaving significant voids or honeycombs internally and on the formed surface. The element shape intricacy will heavily influence the required flow level of a mixture. Figures 7.7, 7.8, and 7.9 show low, medium, and high element shape intricacy.

In a double-tee (Figure 7.7) the concrete flows in a single direction, while in the utility box (Figure 7.8) the concrete is required to flow around corners and cutouts. In a wall with a patterned form liner (Figure 7.9) the concrete is required to flow around corners as well as filling around and into multiple nonuniform spaces. If the intricacy or complexity of the element or structure is high, such as that seen in Figure 7.9, Table 7.10 recommends a high fluidity with a medium to lower viscosity.

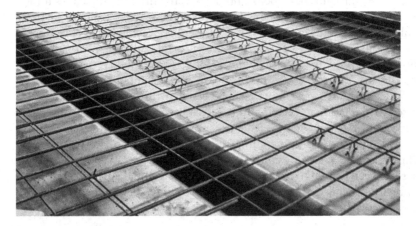

Figure 7.4 Low Reinforcement Level.

Figure 7.5 Medium Reinforcement Level.

Element Depth

A taller element has greater segregation potential than a shallow element such as a slab. If the depth of the element is high, Table 7.11 suggests maintaining a medium to high viscosity to reduce aggregate settlement and bleed. In

Figure 7.6 High Reinforcement Level.

Table 7.9 Fresh Property Guidance Based on Reinforcement Level

		Fresh Properties								
		Fluidity			Passing Ability			Viscosity		
		Low	Medium	High	Low	Medium	High	Low	Medium	High
Reinforcement Level	Low									
	Medium	X			X			X		
	High	X	X		X	X		X		

addition, the practitioner should confirm, through testing, a high level of mixture stability when the element depth is high. Ensuring stability in taller elements becomes even more important with the presence of reinforcing steel or prestressing strand, because of the top bar effect discussed previously in Chapter 4. Figures 7.10, 7.11, and 7.12 provide examples of low, medium, and high element depth.

Figure 7.7 Low Element Shape Intricacy: Prestressed Double-Tees.

Figure 7.8 Medium Element Shape Intricacy: Utility Box.

Figure 7.9 High Element Shape Intricacy: Wall with Patterned Form Liner.

Table 7.10 Fresh Property Guidance Based on Element Shape Intricacy

		Fresh Properties								
		Fluidity			Passing Ability			Viscosity		
		Low	Medium	High	Low	Medium	High	Low	Medium	High
Element	Low									
Shape	Medium	X								X
Intricacy	High	X	X							X

Importance of the Surface Finish

This attribute requires the practitioner to rate the aesthetic requirement of the project. Surface finish becomes more important when the as-cast surfaces will be exposed to public view. Some exposed surfaces may have architectural requirements making them even more important. A low importance would be given to items such as concrete pipe that will be buried underground. ACI

Table 7.11 Fresh Property Guidance Based on Element Depth

		Fresh Properties								
		Fluidity			Passing Ability			Viscosity		
		Low	Medium	High	Low	Medium	High	Low	Medium	High
Element Depth	Low									
	Medium									
	High							X		

Figure 7.10 Low Element Depth: 300 mm Slab.

309.2R presents four classifications of formed surface finishes: Special, Elaborate, Ordinary, and Rough.[41] The author would suggest that in this scenario a High, Medium, or Low rating would correspond to Special, Elaborate, and Ordinary, respectively. Table 7.12 suggests that for higher-quality surface appearance, a medium to high fluidity and a medium to low viscosity will likely provide the best results.

Figure 7.11 Medium Element Depth: Bridge Girder.

Figure 7.12 High Element Depth: Utility Vault.

Table 7.12 Fresh Property Guidance Based on the Importance of the Final Surface Finish

		Fluidity			Passing Ability			Viscosity		
		Low	Medium	High	Low	Medium	High	Low	Medium	High
Importance of Surface Finish	Low									
	Medium	X								X
	High	X	X							X

Element Length

This attribute represents the average distance the concrete must flow from the discharge point. It is a way of considering access level for placement. If access for placement is significantly limited, then the flowing distance will be longer. On the contrary, if the placement equipment can easily move up and down the section while casting, then element length (flowing distance) will be lower. Element length influences both the required flow level and the dynamic stability requirement. For high element length, Table 7.13 suggests a high fluidity to ensure flow over the entire distance and a medium to high viscosity to ensure that the mortar and coarse aggregate do not separate over the long flowing distance. Other proportioning techniques for limiting segregation during long flow include minimizing the coarse aggregate maximum size and minimizing aggregate density.[42]

Table 7.13 Fresh Property Guidance Based on Element Length (or Flowing Distance)

		Fluidity			Passing Ability			Viscosity		
		Low	Medium	High	Low	Medium	High	Low	Medium	High
Element Length	Low									
	Medium	X						X		
	High	X	X					X		

Wall Thickness

This represents the narrowest spacing between two constant surfaces, such as the space between each side of the formwork walls or between the formwork walls and large inserts, etc. As the space decreases, the mixture viscosity becomes more influential. If the viscosity is too high in a narrow space, then the movement through the forms may be slower and can impact placement rate and the final surface finish. A high rating refers to a very open structure with no interference from formed surfaces and therefore no influence on the fresh properties. However, if the thickness is rated as low (meaning it is a narrow and confined opening), Table 7.14 recommends a high fluidity and a medium to lower viscosity.

Coarse Aggregate Content

This consideration is not necessarily an element attribute or characteristic unless the element has an exposed aggregate finish. It is primarily a hardened property requirement that can limit or impact the fresh properties. If certain hardened properties such as modulus of elasticity, shrinkage, or compressive creep are critical, then medium to higher volumes of coarse aggregate may be required, which can impact the achievable slump flow and passing ability. Table 7.15 provides guidance in this sense: for example, if an exposed aggregate finish is required, it will likely be very difficult for this mixture to provide a high level of passing ability without vibration. Although no limit is placed on fluidity, it will be difficult to achieve high flow without segregation when the coarse aggregate level is high.

Concrete Placement and Placement Energy

To the author's knowledge, no systematic study has yet been conducted to quantify the effect of placement technique on SCC performance. However,

Table 7.14 Fresh Property Guidance Based on Wall Thickness

		Fresh Properties								
		Fluidity			Passing Ability			Viscosity		
		Low	Medium	High	Low	Medium	High	Low	Medium	High
Wall	Low	X	X							X
Thickness	Medium	X								
	High									

Table 7.15 Fresh Property Guidance Based on Proportioning with a Higher Coarse Aggregate Volume

		Fresh Properties								
		Fluidity			Passing Ability			Viscosity		
		Low	Medium	High	Low	Medium	High	Low	Medium	High
Coarse	Low									
Aggregate	Medium						X			
Content	High				X	X				X

it is clear through experience that the technique used does influence how SCC flows during placement. Therefore, in the planning process, the parties involved should discuss the proposed technique and consider its impact. The following are some considerations.

CONTINUOUS VERSUS DISCONTINUOUS CASTING

A continuous casting process is one in which, once casting has begun, the entire form is filled without stopping. A discontinuous casting is one that requires multiple deliveries of concrete to the form, with stoppage of concrete placement between successive batches. In this case, the workability retention and the thixotropic characteristics of the mixture must be considered to eliminate the potential for pour lines or other defects.

DROPPING HEIGHT

How high, and through what is the concrete falling? This will affect the required stability of the mixture and should be minimized as much as possible. If it is falling from a high distance through various levels of reinforcement, one should recognize that the stability of the mixture could be compromised. In the case of dropping through reinforcement, one should do everything possible from a placement technique perspective to not have the SCC subject to this situation.

PUMPING CONCRETE

Pumping SCC is quite common. However, one must still assess the pumpability of the SCC mixture, in particular as it relates to mixture stability. If not proportioned correctly, SCC mixtures can segregate in a pump line similar to conventional slump concrete.

PLACEMENT ENERGY

The energy delivered into the concrete during placement is influenced by the technique used. This in turn impacts the fresh properties required for successful placement. Conceptually, this relationship between placement energy and fresh properties is based on the relationship between potential and kinetic energy. In Table 7.16, a high fluidity and a medium to low viscosity is recommended when the placement energy is low. In addition, note that if the placement energy is high, a low-viscosity mixture is not recommended, as segregation may occur.

Combining Tables 7.9–7.16 creates a single table (Table 7.17) used to estimate fresh property targets. The use of this matrix not only assists in choosing initial targets, but also reveals to the practitioner the presence of potentially conflicting requirements.

Table 7.18 provides an example of estimating SCC properties for the production of a precast utility vault using this matrix. The example placement technique is a crane and bucket with multiple deliveries to complete the casting. In this table, the rank for each attribute is presented, as are the fresh property target limitations. For example, the reinforcement level attribute is ranked as medium; with this ranking, a medium to high performance for fluidity, passing ability, and viscosity is recommended. Note that the low performance level for those properties is checked and therefore should be avoided if possible. Including only the chosen or assigned rank for each attribute reveals the initial target level for each fresh property.

In this example, the narrow walls result in a high fluidity level recommendation. A medium passing ability is recommended because of the need to balance the reinforcement level and the coarse aggregate content attributes, and a medium viscosity is also recommended because of the need to balance multiple characteristics.

In order to convert this qualitative analysis to numerical ranges for the fresh properties, Table 7.19 was created. It includes suggested ranges for

Table 7.16 Fresh Property Guidance Based on Placement Energy

		Fresh Properties								
		Fluidity			Passing Ability			Viscosity		
		Low	Medium	High	Low	Medium	High	Low	Medium	High
Placement Energy	Low	X	X							X
	Medium	X								
	High							X		

Table 7.17 Project Attributes and Fresh Property Matrix

| | | Fresh Properties | | | | | | | | |
| | | Fluidity | | | Passing Ability | | | Viscosity | | |
		Low	Medium	High	Low	Medium	High	Low	Medium	High
Reinforce-	Low									
ment	Medium	X			X			X		
Level	High	X	X		X	X		X		
Element	Low									
Shape	Medium	X								X
Intricacy	High	X	X							X
Element	Low									
Depth	Medium									
	High								X	
Importance	Low									
of Surface	Medium	X								X
Finish	High	X	X							X
Element	Low									
Length	Medium	X						X		
	High	X	X					X		
Wall	Low	X	X							X
Thickness	Medium	X								
	High									
Coarse	Low									
Aggregate	Medium					X				
Content	High				X	X				X
Placement	Low	X	X							X
Energy	Medium	X								
	High								X	

(Left margin label: Member Characteristics)

multiple test methods for both passing ability and viscosity, including the U-box, L-box or J-ring, and the T50 or V-funnel time, respectively.

The precast utility vault example can now be converted to actual target values. If one had no previous experience using SCC for this application, and were to use the slump flow, T50, and J-ring tests, then the matrix would suggest the following performance starting points:

Table 7.18 Project Attributes and Fresh Property Matrix (for Precast Utility Vault)

| | | Fresh Properties | | | | | | | | |
| | | Fluidity | | | Passing Ability | | | Viscosity | | |
Member Characteristics		Low	Medium	High	Low	Medium	High	Low	Medium	High
Reinforcement Level	Medium	X			X			X		
Element Shape Intricacy	Medium	X								X
Element Depth	High							X		
Importance of Surface Finish	Medium	X								X
Element Length	Medium	X						X		
Wall Thickness	Low	X	X							X
Coarse Aggregate Content	Medium					X				
Placement Energy	Medium	X								

- Slump flow = 650–800 mm
- T50 time = 2–4 s
- J-ring = 40–80 mm spread difference.

These starting points can then be modified at the practitioner's discretion based on experience.

Table 7.19 Test Method Value Ranges as They Relate to Matrix Performance Levels

	Performance Level		
	Low	Medium	High
Slump Flow (mm)	500–574	575–675	>675
T50 (s)	<2	2–4	>4
V-Funnel Time (s)	<4	4–9	>9
U-Box, obstacle with rising height = min. 300 mm	R3	R2	R1
L-Box, obstacle with h2/h1 = 0.80		2 bars	3 bars
J-Ring*		40–80	<40

*Based on the ASTM J-ring dimensions and expressed as the final spread difference between the traditional slump flow spread and the J-ring spread.

The guidelines reviewed thus far tend to focus on the SCC properties required during the casting process. There are, however, other more subtle properties to consider prior to proportioning, such as those associated with the finishing process. One should consider the intended finish required for a horizontal surface. Is it a broom finish, a hard troweled finish, or a rough finish? Depending on the placement and finishing processes and the timing involved, one may need to adjust the viscosity and the stiffening and setting characteristics of the mixture. For example, a broom finish does not require the concrete to be set, but the concrete does need to be at a stiffer-than-SCC consistency for the broomed ridges to hold. For a final steel-troweled finish, the setting characteristics are more important so that the concrete surface can support laborers or equipment.

Summary

That SCC is more than just a single set of performance properties is evident based on the review of real-world SCC applications presented here. A wide range of performance across three of the main SCC properties of fluidity, passing ability, and viscosity is possible. For concrete practitioners with limited experience either producing or using SCC, the process of selecting the appropriate performance level can be complex. Many times, a less experienced practitioner will choose the highest-performing SCC mixture possible. Technically, this will generally satisfy the performance needs; however, it will typically be at a higher cost. Therefore, the process of performance selection is not only technical in nature but also financial. Balancing performance versus cost is the goal. The guidelines presented are intended to assist in the selection of initial targets. These targets should always be evaluated and confirmed by field trials and mock-ups. Their applicability will be based on the conditions present on a given project or in a given plant. A clear understanding of the anticipated benefits of using SCC will also assist in ranking the importance of specific project attributes, and completing this process equips the practitioner with a clearer set of goals prior to running the first mixture.

Chapter 8

Proportioning SCC Mixtures

Introduction

The process of combining raw materials to create an SCC mixture is not significantly different from that used to develop proportions for conventional concrete. The absolute volume method is still employed to ensure the correct yield of a cubic meter or cubic yard of concrete. Although the general process is similar, there are, however, some concepts relating to the flow and stability properties of the concrete that are new and important to consider. Since the rational mix proportioning method developed by Okamura and Ozawa, numerous SCC mixture proportioning methods have been developed.[1] The purpose of this chapter is not to propose a new methodology, but rather to review the important concepts that form the basis of proportioning SCC mixtures. In this chapter, the following items will be considered: the requirements of an SCC mixture from the perspectives of the producer, contractor, and owner; the roles of paste, mortar, and aggregate in concrete mixtures and how they impact performance specific to SCC; and the boundaries of SCC mixture proportions based on published documents. Finally, supplementary information will be provided to enhance the SCC proportioning techniques and tools already in existence.

A concrete mixture must meet the performance expectations of a wide range of people or groups over its lifetime. From the concrete producer, to the placement crew and contractor, to the owner or owner's representative, each of these groups has a particular perspective regarding which properties are important. The development of the right mixture involves striking a balance between economics, workability, hardened properties, and the ease of reproducing the same performance from batch to batch. What does each of the mentioned groups desire from SCC?

- *Concrete producer.* The concrete producer's desire is to develop a concrete mixture that satisfies a customer's performance requirements, and is easy to reproduce with reasonable effort and cost.
- *Concrete contractor.* A contractor (or placement team in a precast concrete facility) wants a concrete mixture that is easy to place and finish

without delays in setting time while achieving the required hardened properties as designed by the engineer. A contractor also wants each batch of a concrete mixture to be consistent and have the same fresh and hardened properties, all at a reasonable cost.

- *Owner/engineer/architect.* The owner or the owner's representative wants a concrete mixture that produces the final hardened properties required for the structure, including aesthetics and surface appearance, to ensure a long service life for the structure.

This balancing of multiple needs and desires is the essence of mixture proportioning for all concrete, including SCC. To develop mixtures that satisfy all these needs, a more detailed understanding of how mixture constituents interact is useful.

The Concrete Mixture: Breaking It Apart and Putting It Back Together

Of primary importance for all concrete mixtures are the hardened properties. SCC is proportioned to facilitate placement and consolidation, but the fresh property attributes must be secondary to properties such as compressive strength, drying shrinkage, compressive creep, modulus of elasticity, and others. The concrete practitioner must first review the required hardened properties for the project and determine if they place any limitations on the types or amounts of materials used in the SCC mixture. The hardened properties may influence the water/cement (w/c) ratio, water content, coarse aggregate content, paste volume, and powder composition, as well as other factors. Proportioning for the SCC properties must then take these limits into consideration.

The constituent materials used to proportion SCC mixtures are, for the most part, those used for conventional concrete mixtures: Portland cement, fine aggregate, coarse aggregate, water, chemical admixtures, supplementary cementitious materials, and others. One difference is the use of certain non-cementitious fine powders to increase the paste content in SCC. Like conventional concrete, SCC can be evaluated by its building blocks of paste and mortar, which we define below:

- *Paste + air* = cement + other powders (including fine material from aggregate) + water + air
- *Mortar* = paste + fine aggregate
- *Concrete* = mortar + coarse aggregate.

Another valid perspective is to look at concrete as a combination of fluid phase (paste) and solid phase (fine and coarse aggregates). In some respects, concrete is analogous to the human body: the solid aggregates form the skeleton and the paste acts like muscle. The way in which the muscle and skeleton work together determines the strength and mobility of the body. Similarly, how the combined properties of the paste and aggregates interact determines the mobility of concrete and especially SCC. For the remainder of this chapter, SCC proportioning will be considered from the perspective of paste and aggregates (fine and coarse combined) and focus on their fresh property influence.

Considering the Aggregates

When proportioning an SCC mixture, the following aggregate information is used.

Maximum Size Aggregate (MSA)

This characteristic will influence the segregation potential, passing ability, and robustness of a mixture. Larger aggregates have a higher segregation potential and smaller aggregates will enhance the robustness of an SCC mixture.[2,3] When discussing MSA and passing ability, it is appropriate to differentiate between the terms "maximum size" and "nominal maximum size," as they are sometimes erroneously used interchangeably. ACI concrete terminology defines them as follows:[4]

- *Maximum size:* in specifications for and in descriptions of aggregate, the smallest sieve opening through which the *entire* amount of aggregate is *required* to pass.
- *Nominal maximum size:* in specifications for and in descriptions of aggregate, the smallest sieve opening through which the *entire* amount of aggregate is *permitted* to pass; that is, a small percentage of the sample weight may be retained on this sieve.

For example, in an ASTM C 33 #57 grading, the maximum size is 37.5 mm and the nominal maximum size is 25 mm.[5]

When choosing the aggregate size for a conventional slump concrete mixture, ACI 211 states that the largest size possible should be used, as this reduces paste content requirement.[6] This occurs in two ways: it broadens the particle size distribution, allowing for denser particle packing (assuming a well-graded material), and it reduces surface area. ACI 211 also states the following nominal maximum size limitations:

- ≤⅕ of the narrowest dimension between sides of forms
- ≤⅓ of the depth of slabs
- ≤¾ of the minimum clear spacing between individual reinforcing bars, bundles of bars, or pretensioning strands.

Specific Gravity

As in the case of proportioning conventional concrete, the specific gravity is required to convert mass to absolute volume.

Combined Aggregate Void Content

From the local aggregates available, this measurement helps to determine the correct ratio of coarse and fine aggregates. It does not determine the mass used in a mixture, but rather the correct aggregate ratio.[7-9] When blended together, a combination of aggregates will have a certain volume of void space between the particles. The preferred ratio is the one with the lowest void volume, as this void volume must be filled with more expensive paste. Figure 8.1 shows two different aggregate combinations and the resulting void space.

The combined void content is a packing density concept—the broader the particle size distribution, the more densely packed the system will be. The choice of MSA will therefore influence the void volume. The ratio of available aggregates to minimize void content can also be approximated by choosing the ratio that most closely fits an "ideal" grading curve such as those developed by Fuller and Thompson, and others.[10,11]

Particle Shape and Angularity

Rounded, equidimensional aggregates tend to have higher packing density than angular, irregularly shaped ones. This information also provides a relative direction for proportioning. When proportioning with aggregates that are flat

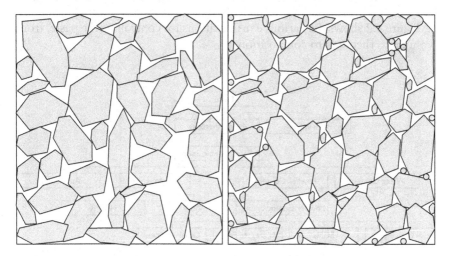

Figure 8.1 Schematic Examples of Void Space with Two Different Aggregate Combinations.

and elongated, relatively more paste will be required to achieve flow than when using aggregates with more equidimensional particles. Similarly, aggregates with greater angularity require more paste than rounded aggregates. Rounded equidimensional particles also promote better mobility.[12]

Particle Size Distribution/Gradation

The overall gradation of the aggregate skeleton provides an indication of total aggregate surface area. Knowing that the surface area is relatively higher or lower provides a direction for initial proportions as well as trial adjustments. For stability purposes, being aware of the quantity of material passing the 300 μm and 75 μm sieves will influence the proportion of cement or other powder included in the mixture.

Considering the Paste/Fluid Phase

In this book, we consider the paste to be composed of water, air, and all powder material finer than the 125 μm sieve. The paste is critical to SCC performance: without sufficient, properly proportioned paste, the SCC mixture will not perform as desired, regardless of how well graded, round, and clean the aggregates may be. When proportioning SCC, two main considerations are given with respect to the paste fraction:

1. How much: what is the paste volume to be used?
2. What is the paste rheology: how does the paste flow and how easily do the solid materials move within it to avoid segregation and blocking?

Figure 8.2 shows the various considerations for controlling the paste, from volume to rheology to composition.

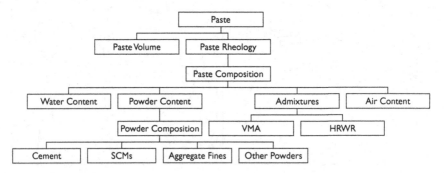

Figure 8.2 Items to Consider when Proportioning the Paste Fraction/Fluid Phase in SCC Mixtures.

Paste Volume

The excess paste theory suggests that in order to enhance concrete flow, not only do the void spaces between the solid particles need to be filled and the surface area covered, but an additional "excess" amount of paste is necessary to further separate the solid particles (Figure 8.3).[13] With a constant paste composition, as the paste volume and the thickness of the paste layer around each particle increase, the likelihood of aggregate particle collision decreases, thereby increasing the slump flow and decreasing the viscosity of the concrete mixture.[14,15]

Paste volume is one of the most important controlling factors in proportioning a well-performing SCC mixture. It is really quite simple: the higher the paste content, the more like paste the concrete will flow, with less chance for blocking and segregation. The real challenge is to balance this fresh property improvement with hardened property requirements.

Paste Rheology

Controlling paste rheology is more complicated than controlling paste volume. Numerous tools are at the practitioner's disposal for controlling paste rheology, including water, high-range water reducers (HRWRs), viscosity-modifying admixtures (VMAs), cement, supplementary cementing materials (SCMs), and other fine powders. All concrete practitioners are aware of the w/c ratio and the water-to-cementitious materials ratio, each calculated by mass and used for controlling properties such as compressive strength and durability. Most concrete practitioners understand that, all other things being equal, as the w/c ratio decreases, the concrete tends to become more cohesive or viscous. Many of the powders used for proportioning SCC, however, have different specific gravities, making a mass calculation less useful for comparing mixtures and controlling rheology. A ratio that is useful for controlling the paste rheology, therefore, is the water-to-powder ratio by volume, wv/pv.[16] The original proportioning method developed by Okamura

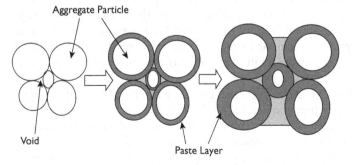

Figure 8.3 Excess paste concept.

suggests a wv/pv ratio between 0.9 and 1.0, which would correspond to a w/c ratio between 0.29 and 0.32, assuming only Portland cement was used. Based on practical experience, these mixtures would be fairly "sticky," meaning their viscosity would be high.

Chemical admixtures such as HRWRs and VMAs are used to control the paste rheology and therefore the concrete rheology without the need to adjust mixture proportions. For example, HRWRs are used to increase fluidity without adding water, which can reduce viscosity. VMAs are used to increase viscosity while neither decreasing water nor increasing powder.

In addition to adjusting the content of water or powder and therefore manipulating the wv/pv ratio, or using admixtures, the powder composition can be adjusted to influence the paste rheology.[17,18]

POWDER COMPOSITION

The powder composition will influence all aspects of the SCC mixture, from the fresh properties and rheology, and the stiffening and setting characteristics, to the hardened properties, as well as to the effectiveness and dosage requirements of HRWR admixture. Choosing the right powder composition requires knowledge of local materials and then deciding on the ratio of those materials. The choice of powders, and their ratios, will be dictated by their effect on performance, their availability, cost, and the ability of the production plant to handle multiple powders.

The starting point for powder composition is the Portland cement. The cement chosen when proportioning SCC has a great influence on the paste rheology, as cements can vary in water demand and reactivity to HRWR chemistries.[19] For example, if a mixture is initially proportioned with a particular cement, and a new cement, with a lower water requirement, is substituted for the original, there is a good chance that either the water content or the HRWR dosage will change. This adjustment, however, can change the paste rheology, especially if the water content is reduced. Because of the variability in characteristics within and between different powder types, no single statement can capture the influence a powder has on fresh SCC properties. If a powder is used to replace cement, the influence will always be relative to the characteristics of that particular cement mixture. Because cement characteristics (such as fineness) change, the relative magnitude of the impact of other powders will vary. For this reason, when new powders are being considered, testing must be done.

Tables 8.1–8.3 provide very general guidance regarding the impact of fly ash, slag, and silica fume on certain fresh properties of SCC mixtures. The statements contained in these tables assume the powders are replacing a normal ASTM C 150 Type I cement by mass at normal levels.

Other powders besides those mentioned above are used in the production of SCC, including, but not limited to, milled or ground limestone powder,

Table 8.1 Impact of Fly Ash Replacement on Fresh SCC Properties

Fluidity/Filling Ability	Depending on the specific gravity of the fly ash and the replacement level, both the paste volume and the wv/pv ratio are influenced. The paste volume will increase and the wv/pv ratio will decrease, having opposite effects on mixture fluidity. Within normal replacement levels, the increased paste content combined with the spherical particle shape will tend to increase the fluidity of an SCC mixture, even though water content stays constant.[20,21] This will be the primary effect as long as the wv/pv ratio is not reduced too much.
Passing Ability	The increase in paste volume and subsequent decrease in aggregate content will lead to an improvement in the passing ability of the mixture.[16] The degree to which it improves is dependent upon the specific gravity of the fly ash being used and the replacement level.
Segregation Resistance	The increase in powder volume by replacing cement with fly ash leads to a lower wv/pv ratio, as well as a greater surface area. This will lead to a reduction in the bleeding potential of the mixture.[20] The lower wv/pv ratio results in a more cohesive mixture and a reduced tendency for aggregate settlement. If the fly ash has higher fineness than the cement, bleeding and segregation will decrease.

Table 8.2 Impact of Slag Cement Replacement on Fresh SCC Properties

Fluidity/Filling Ability	The specific gravity of slag cement is usually lower than that of cement but higher than that of fly ash; therefore, a replacement of cement by mass does not increase the paste volume to the degree seen with fly ash. Research has shown, however, that the replacement of cement with slag cement improves the fluidity of paste and mortars, thereby decreasing the water or HRWR requirement for a set workability.[22]
Passing Ability	The increase in paste volume is lower than when fly ash is used, and therefore, depending on the replacement level, the paste volume increase may be minimal and no effect on passing ability be seen.
Segregation Resistance	The impact on bleed is determined by the particle fineness relative to the cement. As the slag becomes finer than cement, bleed will be decreased; if it is coarser than cement, bleeding may increase.[22]

Table 8.3 Impact of Silica Fume Replacement on Fresh SCC Properties

Fluidity/Filling Ability	In some cases, the smaller particle size leads to improved packing density of the powders, resulting in a more fluid paste. There is a limitation to the packing density benefit, however. If too much silica fume is added, the surface area increase may become dominant, thereby increasing water and/or HRWR demand.[23,24] This behavior has also been reported to be dependent on the HRWR chemistry being used.[25]
Passing Ability	Typically, the amount of silica fume used will be less than that of fly ash or GGBFS,* and therefore the increase in paste volume and subsequent decrease in aggregate volume will not be significant enough to change the passing ability.
Segregation Resistance	As the silica fume content increases, the bleeding and segregation tendencies of concrete decrease.[23]

* GGBFS = ground-granulated blast-furnace slag

aggregate fines, and precipitated silica. Selecting the correct paste and powder composition can be the most intensive part of proportioning an SCC mixture with respect to testing, assuming that the producer has numerous powders available.

Existing SCC Proportioning Methodologies

Over 15 SCC mixture proportioning procedures have been developed worldwide.[1] Of these, only two include compressive strength as part of the criteria, while most methods proportion to achieve only the fresh SCC properties.[1,26] The techniques that focus on the fresh properties can be divided into two basic groups:

1. Those based on calculated values derived from testing and evaluation of the raw materials intended for use.
2. Those based on choosing aggregate, powder, and water amounts from a series of general tables.

Regardless of the process, all methods require testing of the proportions selected. No method exists that will provide the optimum mixture simply by analysis of the material characteristics. What the methodologies do, however, is to reduce the number of trials required. This section provides the basic steps of four selected methodologies. The intent is to understand the logic behind proportioning SCC mixtures and the process for decision-making. The details of these methods can be found in the references cited.

The Rational Mix Design Method

In the rational mix design method presented by Okamura and Ozawa, the following four steps are followed:[16]

1. Coarse aggregate content is fixed at 50% of the solid volume.
2. Fine aggregate content is fixed at 40% of the mortar volume.
3. The wv/pv ratio is assumed to be between 0.9 and 1.0, depending on the properties of the powder.
4. Superplasticizer dosage and the final wv/pv ratio are determined so as to ensure self-compactability.

The CBI Method

In the CBI method developed by Billberg et al. the following three steps are taken:[15]

1. Calculate the minimum paste volume according to aggregate properties, hardened property criteria, and passing ability criteria (structural detailing). Calculations for determining coarse aggregate amount based on the relationship between MSA and the smallest gap through which the concrete must flow are provided. A methodology for establishing the paste volume considering the above information and the void content of the aggregate skeleton is included.
2. Design the paste based on rheological measurements of the fine mortar phase.
3. Test and verify the desired fresh and hardened concrete properties.

The ICAR Method

The method developed by the International Center for Aggregates Research suggests the following three steps:[8]

1. Select the aggregates and MSA to be used and evaluate them for packing density, shape, and angularity, and determine the void content for the various blends chosen.
2. Determine the appropriate paste volume for the desired filling and passing ability based on the void content, the shape and angularity assessment from Step 1, and robustness requirements.
3. Determine the paste composition based on hardened and fresh property requirements.

The ACI 237 Method

ACI 237R-07 provides the following guidelines for proportioning SCC mixtures (a number of tables accompany these steps, but are not reproduced here):[21]

1. Determine the appropriate slump flow requirements based on the application.
2. Select the appropriate maximum size coarse aggregate and coarse aggregate content based on the passing ability requirements. The method proposes a starting coarse aggregate content of 50% of the absolute volume of the mixture. This is based on the rational mix design methodology mentioned above.
3. Estimate the required cementitious/powder content based on the slump flow requirement.
4. Calculate the paste and mortar volumes.
5. Select the appropriate admixtures for use.
6. Batch trial mixtures.
7. Test the mixture for the required fresh and hardened properties.
8. Based on the results, adjust the proportions, re-batch, and test again.

Whichever method for proportioning SCC is followed, it should include consideration of hardened properties and provide guidance for achieving the necessary fresh properties. In some cases, passing ability will be critical, and therefore choosing the coarse aggregate size and volume will be the first priority. If passing ability is not critical, other performance requirements, such as surface finish importance, may dominate the process, and these requirements should be clear from the start.

SCC Proportioning Guidance

The following points were created as a supplement to existing proportioning methods. Tables 8.4–8.8 are intended to assist the practitioner in making decisions regarding initial mixture proportions. Some are based on data developed through the internal case study analysis and others through Domone's case study.[49] In the author's case study analysis, the paste volume and wv/pv ratio were estimated using the following specific gravities for powders when they were not provided: fly ash 2.4, limestone powder 2.7, ground-granulated blast-furnace slag (GGBFS) 2.9, and silica fume 2.3. Table 8.4 presents the case study mixture proportioning data used.

Choose the Maximum Aggregate Size

This is intended primarily to ensure adequate passing ability, although, additionally, larger aggregates have a greater tendency to segregate.[50] The choice can be based on prior experience with the element being cast or, if no such experience exists, the following ACI recommendations for nominal maximum size limitations can be used:

- For forms, ≤⅕ of the narrowest dimension between sides.
- For slabs, ≤⅓ of the depth.

- For individual reinforcing bars, bundles of bars, or pretensioning strands, ≤¾ of the minimum clear spacing between them. To ensure passing ability, the author suggests that if no prior experience exists and passing ability is a critical performance characteristic, this be reduced to ½ of the minimum clearance. This should also include clearance between reinforcement and the sides of forms.

EXAMPLE

A large precast/prestress double-tee producer in North America successfully uses SCC with a slump flow of 500–550 mm. In this application, two pre-stressing strands run parallel along the length of the stem. The clear spacing between the strands is 31.75 mm. The clear spacing between the strands and the form is 76 mm. On the deck, the clear spacing between the wire mesh and the deck form is 38 mm. This producer uses a coarse aggregate with a nominal maximum size of 19 mm and a maximum size of 25 mm.

Coarse aggregates from 10 to 40 mm have been used in producing SCC mixtures.[49]

Determine the Appropriate Ratio of Fine, Intermediate, and Coarse Aggregates

One of two methods can be used. Some methods suggest choosing the blend of all aggregates and others suggest choosing a coarse aggregate content.

CHOOSING THE BLEND

The optimum ratio of aggregates will minimize the void content of the combined aggregate skeleton. This combination can be determined by measuring the combined aggregate void contents[8,15] or fitting the overall combined gradation to an "ideal" grading curve such as the power 0.45 curve, Bolomey's curve, or others.[8,11] Passing ability requirements will also influence the aggregate blend used. The CBI method contains an equation for calculating coarse aggregate content based on aggregate size.

CHOOSING A COARSE AGGREGATE CONTENT

This method may be preferred when certain hardened properties such as modulus of elasticity are critical. Table 8.5 shows the coarse aggregate volumes based on aggregate size from Domone's analysis. A wide range of coarse aggregate volumes have been used in practice. The aggregate volume is higher and the range is wider for larger MSA.

Table 8.4 Mixture Proportions Case Studies

Ref.	Faster Speed of Construction and Time Savings	Labor Savings	Ease of Placement and Consolidation Through Dense Reinforcement	Ease of Placement and Consolidation in a Complex Structure or Shape	Ease of Placement with Limited Access	Provide Higher In-Place Quality and Aesthetics	Improved Worker Safety and Noise Reduction	Cement (kg/m^3)	Other Powder (kg/m^3)	Other Powder Type
27				X	X			533	0	Limestone powder
27	X	X			X		X	306	319	Limestone powder
27			X					302	333	Slag cement
27								340	215	Slag cement
28	X	X						515	70	Limestone powder
29						X	X	415	180	Limestone powder
29						X	X	405	121	Limestone powder
30	X					X	X	367	150	Limestone powder
21				X	X			309	77	Slag cement and fly ash
21			X	X	X	X		230	155	Slag cement
21		X				X		385	68	Slag cement
21								445	0	Fly ash
31					X			400	100	Limestone powder
32			X		X			330	270	Limestone powder
32			X		X			314	206	Limestone powder
33					X			438	70	Silica fume
34					X		X	415	16	Slag cement
35						X		315	135	Limestone powder
36			X					370	240	Limestone powder
36					X			370	240	Limestone powder
36					X			300	200	Slag cement

								Material
37			X			380	95	Slag cement
38						395	70	Slag cement
39		X		X		400	150	Fly ash
39	X	X		X		355	310	Fly ash
40				X		435	75	Limestone powder
41	X			X		400	100	Limestone powder
41	X	X		X		465	65	Silica fume
42	X	X		X		480		Fly ash
43		X		X		364	186	Filler
44		X		X		364	211	Limestone powder
45			X			265	200	Limestone powder
46		X	X	X		335	110	Fly ash
47						356	119	Slag cement
47						405	47	C Fly ash
47						415	83	C Fly ash
47						475	0	
47						256	178	Slag cement
47						415	0	
47						403	0	
47						445	0	
21						356	148	Fly ash
21						403	0	
21						368	107	Slag cement
21						205	268	Slag cement and fly ash
21						415	0	
21						356	59	Fly ash
48	X				X	455	70	Limestone powder

Continued

Table 8.4 Continued

Ref.	Specific Gravity*	Total Powder (kg/m³)	Coarse Aggregate (kg/m³)	Fine Aggregate (kg/m³)	Water (kg/m³)	Air (%)	VMA	wv/pv	Paste Volume (%)	Paste Volume (%) Minus Air
27		533	871	773	160	4.5	Yes	0.94	37.5	33.0
27	2.7	625	864	664	168	4.5	Yes	0.78	42.9	38.4
27	2.7	635	884	692	166	2.0	Yes	0.76	40.6	38.6
27	2.94	555	822	786	175	4.5	Yes	0.96	40.1	35.6
28	2.7	585	789	737	175	4.5	No	0.92	41.0	36.5
29	2.7	595	697	852	166	6.2	No	0.83	42.7	36.5
29	2.7	526	733	895	162	5.0	No	0.93	38.6	33.6
30	2.7		821	804	165	4.5	Yes	0.96	38.2	33.7
21	2.6	386	1006	758	166	6.0	No	1.30	35.4	29.4
21	2.9	385			180	6.0	Yes	1.42	36.7	30.7
21	2.9	453			181	5.5	Yes	1.24	38.2	32.7
21					178	6.0	Yes	1.26	38.0	32.0
31	2.4	500	890	890	162	4.5	Yes	0.96	37.6	33.1
32	2.7	600	844	839	165	2.0	Yes	0.80	39.0	37.0
32	2.7	520			157	2.0	Yes	0.89	35.3	33.3
33	2.7	508	792	809	180	4.0	Yes	1.09	38.5	34.5
34	2.3	431	700	931	196	5.0	Yes	1.41	38.5	33.5
35	2.9	450	900	825	190	2.0	Yes	1.29	35.7	33.7
36	2.74	610			171	2.0	No	0.83	39.6	37.6
36	2.74	610			171	2.0	No	0.83	39.6	37.6
36	2.91	500			170	2.0	No	1.03	35.4	33.4
37	2.9	475			171	6.0	No	1.11	38.5	32.5
38	2.9	465	750	1040	195	2.0	No	1.30	36.5	34.5

	Sp. gr.*									
39	2.4	550	764	760	190	2.0	Yes	1.00	40.0	38.0
39	2.4	665	704	704	167	2.0	No	0.69	42.9	40.9
40	2.7	510	703	967	194	2.0	Yes	1.17	38.0	36.0
41	2.7	500	875	870	180	2.0	Yes	1.09	36.4	34.4
41	2.3	530	915	710	175	2.0	No	0.99	37.1	35.1
42	2.4				178	6.5	Yes	1.16	39.6	33.1
43	2.7	550	700	970	200	2.0	No	1.08	40.5	38.5
44	2.4	575	830	805	172	6.0	No	0.89	42.6	36.6
45	2.6	465	682	934	190	2.0	No	1.13	37.8	35.8
46	2.9	445	615	1048	170	2.0	Yes	1.14	31.9	29.9
47	2.4	475	984	764	180	6.0	No	1.17	39.4	33.4
47	2.4	452	950	840	174	6.0	No	1.17	38.3	32.3
47		498	1195	686	178	2.0	Yes	1.07	34.5	32.5
47	2.9	475	890	1012	180	2.0	No	1.19	33.1	31.1
47		434	920	977	148	2.0	Yes	1.04	29.1	27.1
47		415	890	981	154	2.0	Yes	1.17	28.6	26.6
47	2.4	403			162	2.0	Yes	1.26	29.0	27.0
21		445			157	2.0	Yes	1.11	29.9	27.9
21	2.9	504			186	6.0	No	1.06	42.1	36.1
21	2.6	403			178	6.0	No	1.39	36.7	30.7
21		474			185	6.0	No	1.20	39.9	33.9
21	2.4	473			161	6.0	Yes	0.96	38.9	32.9
48	2.7	415			170	6.0	Yes	1.29	36.2	30.2
		415			166	6.0	Yes	1.20	36.4	30.4
			730		185	1.0	Yes	1.08	36.6	35.6

*Figures in shaded cells are estimated specific gravities.

Table 8.5 Coarse Aggregate Volume per Cubic Meter versus MSA[49]

	Average	Median	Range
<20 mm	31.0%	30.9%	28.3–34.9%
≥20 mm	32.3%	31.7%	28–42.3%

Determine the Quantity of Paste Required

The void content of the aggregate skeleton, the surface area and particle shape of the aggregates, and the slump flow target all influence the necessary paste volume. Table 8.6 provides information from the author's and from Domone's case study analyses. Included are paste volume ranges from the author's analysis calculated both to include and not to include the air content. In some cases, air contents were not reported. In these cases, the author assumed an entrapped air content of 2%.

When compared with the range of paste volumes without air, the paste volume with air does not increase the maximum amount significantly. It does, however, raise the minimum value by 6% and narrow the overall range from 12% to 8%.

As previously discussed, the necessary paste volume will fluctuate with a number of factors, including the aggregate void content. The void content is driven by packing density, which is influenced by the maximum aggregate size. In Domone's case study analysis, higher median paste contents were observed for mixtures using smaller maximum size coarse aggregate. Paste requirement is also influenced by particle shape. In that same analysis, higher paste requirement was seen with crushed aggregates. These results are not unexpected and provide another example of how the rules of thumb for SCC mixture proportions are similar to those of conventional concrete.

As slump flow increases, higher paste contents are recommended. ACI 237 suggests this trend by recommending higher powder contents for higher slump flow targets. If an SCC mixture is proportioned with insufficient paste, the practitioner's tendency will be to increase the HRWR dosage. This, however, can over-disperse the paste and lead to bleeding and instability.

Table 8.6 Paste Volume Data from Case Study Analysis

	Mean	Median	Range
Author's with air	39%	38%	35–43%
Author's without air	35%	34%	29–41%
Domone's	35%	35%	30–42%

Choose the Starting Water Content

Water contents in the case studies analyzed range from 148 to 200 kg/m³, with an average of 174 kg/m³. As a primary component of the paste fraction, the water content has a significant impact on the paste and therefore also the concrete viscosity. The viscosity of an SCC mixture can influence the final benefits achieved on a given project. In some cases, then, the water content can be chosen to achieve a viscosity range and in order to target certain desired benefits. Table 8.7 shows the frequency with which a mixture water content was used for projects with the listed application requirements from the author's case study analysis.

For densely reinforced elements, the water content tends to be lower. This is logical, as the viscosity of the paste is critical in order to avoid coarse aggregate blocking. It is also possible that this trend is influenced by the project's compressive strength requirements. When surface finish is critical, the water content is skewed toward the upper end of the range, which is logical as the viscosity of the paste is preferably moderate to low in order to achieve smooth, bughole-free surfaces. When limited access and complex shape application requirements are critical, the water contents are more evenly distributed.

Choose the w/c Ratio and Air Content Based on Compressive Strength and Durability Requirements

In most cases, the compressive strength of an SCC mixture will be higher than that of a conventional slump mixture at the same w/c ratio (see Chapter 4). However, the starting point can be chosen based on previous experience with the available materials.

Determine Final Powder Composition and Verify the wv/pv Ratio Based on the Fluidity and Viscosity Performance Requirements

Presumably, the paste volume, water content, and maximum w/c ratio necessary for the project have already been determined. From this information, the water and cement volumes can also be determined. If other powders are available, these materials should be added in the amount required to achieve

Table 8.7 Water Content Levels by Application Requirements

Water Content (kg/m³)	<170	170–190	>190
Dense Reinforcement	33%	67%	0%
Surface Finish	0%	82%	18%
Complex Structure	20%	60%	20%
Limited Access	39%	50%	11%

the final paste volume. If no other powders are available, the cement and water contents will have to be increased, at the same ratio, in order to reach the desired paste volume.

The viscosity of a concrete mixture can be changed by adjusting either the volume or the viscosity of the paste. Figure 8.4 shows the relationship between the wv/pv ratio and the paste volume, both including and excluding air, for the case studies reviewed by the author. The lower paste volumes tended to have higher wv/pv ratios, presumably in an effort to maintain a reasonable concrete mixture viscosity. When attempting to keep SCC mixture viscosity lower, the preferred method is to increase the wv/pv ratio rather than the total paste volume, if possible, for cost control reasons. However, this must be balanced against the hardened property requirements.

Table 8.8 presents wv/pv ratios based on the application requirements from the case study analysis. Projects with dense reinforcement fall close to or into the original range suggested by Okamura and Ozawa of 0.9–1.0. Other requirements such as surface finish typically incorporate a relatively higher wv/pv ratio.

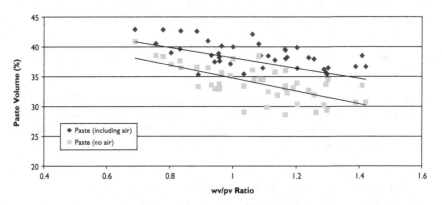

Figure 8.4 Paste Volume as a Function of wv/pv Ratio for Case Studies.

Table 8.8 wv/pv Ratios by Application Requirements

	Average	Median
Dense Reinforcement	1.02	0.99
Surface Finish	1.09	1.11
Complex Structure	1.10	1.08
Limited Access	1.07	1.06

Summary

Proportioning SCC mixtures is not drastically different from proportioning conventional slump mixtures. In the most basic terms, it is the paste fraction (including air) of the SCC mixture that controls the flowing and stability properties of the mixture. The paste also happens to heavily influence numerous hardened properties and the overall mixture cost. In controlling the paste, one will modify the paste volume and/or the paste composition. Optimizing the aggregate skeleton, from particle shape (if possible) to particle size distribution and packing density, will assist in keeping paste contents and costs relatively low.

Numerous SCC proportioning procedures have been developed and provide good direction for choosing initial proportions. As with all mixture proportioning procedures, whether for conventional slump concrete or for SCC, testing and evaluation of the chosen proportions are required. Then, as necessary, adjustments to proportions should be made to achieve the desired final performance. The guidance and tables provided in this chapter are to assist in choosing the initial proportions and in making adjustments.

Chapter 9

Mixture Qualification

Introduction

Once a set of performance targets and mixture proportions have been developed on paper, the next step is to qualify the mixture to ensure successful performance during production. The Interim Guidelines developed by the Precast/Prestressed Concrete Institute (PCI) present three phases in the mixture qualification process:[1] first, small, well-controlled laboratory batches are tested; second, the chosen mixture(s) is tested in production, which includes the use of batching, mixing, and placement equipment; and finally, a quality assurance/control plan for the fresh and hardened properties is developed. This chapter addresses the laboratory testing and production trial phase of the qualification process; quality control will be addressed in Chapter 11.

Laboratory Qualification

A certain amount of testing always follows the development of a new mixture. In the qualification process, one tests both the critical fresh and hardened properties. To ensure the highest level of control, it is recommended that, when possible, small-scale laboratory batches are run first. In this way, the effects of specific and controlled adjustments are learned. When performing concrete laboratory experiments, good practices and all appropriate standards, such as those found in ASTM, should be followed. Figure 9.1 shows a flow chart for the first phase of the qualification process. The mixture developed is tested for the targeted fresh properties. If the properties are not achieved, adjustments to the mixture proportions are made. If the fresh targets are achieved, then testing for mixture robustness and for the required hardened properties is conducted. If the required hardened properties are not achieved, mixture proportion adjustments are made. The cycle continues until the appropriate fresh and hardened properties are achieved.

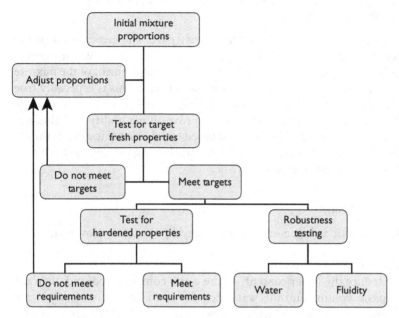

Figure 9.1 Phase One of the Mixture Qualification Process.

Testing Initial Proportions

The scope of the initial laboratory testing program is situation-dependent. It can be as limited or as expansive as desired. In many cases, the concrete practitioner approaches this phase with the sole intent of "producing the mixture" that will be used. This sometimes leads to a narrow, limited view of the process where a single mixture is tested, adjustments made, and the mixture is tested again. This methodology is not wrong, but neither is it efficient or most conducive to learning the impact of multiple variables and their interactions. Efficient experimental design is a discipline in and of itself, and numerous books can be found explaining the topic.[2,3] Statistically designed experimentation can simultaneously develop a specific SCC mixture and highlight interactions between mixture proportioning variables.[4,5] Developing a mixture and understanding interactions are important for both the short and long term. It is the author's opinion that more time invested at this stage, to further one's understanding, results in less troubleshooting time spent in later stages. If the practitioner does not possess experimental design expertise, there are software packages commercially available or services available through materials suppliers that can assist in developing an effective SCC experiment if desired.

Laboratory Testing, Interpreting Results, and Adjusting Mixture Proportions

Testing and producing data is often the easy part. The critical part of the process is interpretation of the data and making adjustments. When it comes to adjusting SCC mixture proportions, a qualitative assessment of the mixture is often as important as any test measurement taken. This is especially true with respect to segregation and instability. The following subsections provide guidance for adjusting proportions to improve filling ability and fluidity, passing ability, stability, and robustness based on laboratory testing. It should be noted that, in some cases, adjustments to the proportions may have an impact on more than one characteristic.

Filling Ability/Fluidity

This characteristic will generally be measured by the slump flow test:[6]

- If the slump flow is too high:
  - o Reduce the water content—if the water content is reduced, then the paste volume and the water/powder ratio by volume (wv/pv) are reduced, both of which will tend to increase the mixture viscosity. This will also require an increase in aggregate content to make up for the lost water volume. A reduction in paste volume can also negatively impact passing ability.
  - o Reduce the high-range water-reducer (HRWR) dosage—this has the benefit of reducing fluidity while not requiring a volume-balancing adjustment to the mixture proportions.
- To help with the decision on which step to take, one should note whether or not the mixture is showing signs of segregation and instability. If the mixture is not showing signs of instability at this higher slump flow, this is a good thing, and the HRWR dosage should be reduced. The subsection below on adjusting to improve stability and segregation resistance will discuss how to address slump flow level if the mixture is showing signs of instability.
- If the slump flow is too low:
  - o Increase the water content as long as this does not increase the water/cement (w/c) ratio past the specified limit. An increase in water will increase the paste volume and the wv/pv ratio, thereby decreasing mixture viscosity as well as increasing fluidity. This type of adjustment should therefore be made in small steps so as to not induce mixture instability.
  - o Increase the HRWR dosage—if slump flow increases systematically with increasing HRWR dosage and without segregation, the mixture proportions are reasonably balanced. If during this process the slump

flow does not increase, this can be due to one or more of the following: the mixture has an insufficient paste volume, the cement used is stiffening rapidly, or the w/c ratio is so low that the dispersant saturation dosage has been exceeded. If this happens, therefore, adjustments to the mixture proportions may be necessary.

o Increase the total paste volume—if the paste volume is too low and one continues to add HRWR, this can cause chemically induced segregation, which is recognizable by a slow bleed and discoloration on the surface of the concrete. See Figure 9.2: notice the lighter colored bleed on the surface and around the edge of the slump flow paddy. An insufficient paste volume is also recognizable by the presence of a concentration of aggregate in the center of the slump flow patty when the slump flow test is performed. This may also be accompanied by the discolored bleed mentioned earlier. When the paste volume is increased, the recommendation is to make this adjustment by increasing both the powder and water contents in order to maintain the wv/pv ratio.

Passing Ability

ASTM C 1621 assesses passing ability of SCC by determining the difference between the slump flow with and without the J-ring.[7] Some European standards require the height difference of the paddy inside the ring versus outside the ring be measured as well.[8,9] This height difference is intended to assess

Figure 9.2 Discolored Bleed due to HRWR Overdose and Insufficient Paste Volume.

whether or not there is significant aggregate bridging and therefore build-up on the inside of the ring.

Using the methodology proposed by ASTM C 1621, if the passing ability is inadequate—meaning that the slump flow *without* the ring is 50 mm (2 inches) or more, greater than the J-ring slump flow—one must again make an assessment as to how the blocking or bridging of aggregates is occurring. Figure 9.3 provides guidance in assessing the passing ability through the J-ring in order to make the appropriate adjustments:

1. J-Ring Slump Flow 1 in Figure 9.3 represents an acceptable result.
2. J-Ring Slump Flow 2 represents a result where the measured slump flow is acceptable, if measured to the edge of the entire flow, but the concrete is separating, as evidenced by a band, or halo, of mortar or paste around the edge of the patty. In this situation, where the mixture is separating, more than one adjustment may be necessary: the first step is to improve mixture stability, then decide whether an adjustment to the coarse aggregate content or maximum size is necessary. The presence of the halo can mean either the paste is not viscous enough (too much water) or is

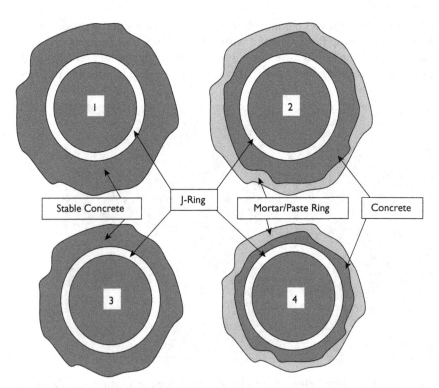

Figure 9.3 Assessment of Passing Ability through J-Ring.

too highly dispersed with HRWR to carry the aggregate through the reinforcement. In the case where the paste is overly dispersed, one may see evidence of this during the conventional slump flow test or during the static stability testing. Signs of HRWR overdose can be significant bleeding and discoloration of the bleed water and/or significant foaming on the top surface of concrete in a wheelbarrow or mixer. This is usually an indication that the paste volume in the mixture is inadequate and should be increased. If it appears that the paste is not viscous enough, addition of a viscosity-modifying admixture (VMA) and/or a reduction in the wv/pv ratio may be necessary. Visual assessment of the severity of aggregate blocking can provide qualitative indications that the coarse aggregate content or maximum size should be reduced. In these instances, taking the measurement of the paddy height on the inside and outside of the ring is also useful for quantifying the performance and making this determination.

3. J-Ring Slump Flow 3 represents a mixture that shows no signs of paste or mortar separation, but the resulting slump flow measurement is too short. If the conventional slump flow achieved its target, this indicates that either the maximum size and/or the quantity of coarse aggregate is too high. Assuming the maximum size was chosen correctly, then the coarse aggregate quantity should be reduced. This can be done either by adjusting the ratio of coarse aggregate to fine aggregate and leaving the paste content the same or by reducing the total aggregate content and increasing the paste content. The former adjustment will influence the fluidity and stability characteristics of the mixture. The latter adjustment will typically be more effective, but is the more costly option. It is also possible to make slight adjustments using both of the options simultaneously.

4. J-Ring Slump Flow 4 is the worst-case scenario, because the mixture is neither stable nor fluid enough. In these cases, there is a lack of paste required to promote the appropriate fluidity and passing ability and the paste is over-dispersed via the use of HRWR. The first adjustment in this situation is to increase the paste volume, re-test, and re-assess the performance.

Stability and Segregation Resistance

The following are steps for adjusting stability based on the bleeding test[10] and the column segregation test:[11]

- Too much bleed water accumulation:
  - o Add or increase the dosage of a VMA.
  - o Lower the wv/pv ratio by either reducing water or increasing powder. If the paste volume is relatively low, then one should increase the

powder content; if the paste volume is high enough then one can reduce the water content. Remember that the overall paste volume is a primary variable in achieving all SCC fresh properties.

o   Increase the fine aggregate content so as to introduce additional fine material (<300 µm).

o   If the excess bleed is chemically induced by too high an HRWR dosage, then lower the HRWR dosage and increase the paste volume. Figure 9.1 shows this in the slump flow paddy, and Figure 9.4 shows this type of extreme bleeding in a laboratory mixer at rest.

- Too much coarse aggregate settlement:
  o   If possible, based on the application requirements, reduce the slump flow by reducing HRWR dosage.
  o   Increase the mixture viscosity by decreasing wv/pv or by adding a VMA or increasing the VMA dosage.
  o   If accompanied by chemically induced bleeding, then increase the paste volume.
  o   If possible, reduce the maximum aggregate size—if this adjustment is made, note that an increase in paste volume may also be required.

Robustness Testing

Once a mixture or set of mixtures that provide the desired fresh and hardened properties have been developed, the robustness of these mixtures should be

Figure 9.4 Extreme Segregation and Bleed in a Mixer at Rest.

determined. Robustness of SCC is defined as the insensitivity of an SCC mixture to fluctuations in fluidity, batching accuracy, and/or raw material characteristics such as moisture content and particle size distribution that could lead to changes in the SCC mixture's fresh properties. A robust mixture is one that can be produced easily and can consistently achieve the targeted fresh properties without continuous involvement of the producer or contractor at the jobsite. In previous chapters, the relationship between increasing fluidity resulting in reduced mixture stability for a given set of materials and mixture proportions was presented. Similarly, others have found that as the slump flow level increases, mixtures become more sensitive to changes and robustness decreases.[12] The goal in testing robustness is to find the point at which a mixture will no longer exhibit acceptable performance in terms of stability, fluidity, and passing ability. Therefore, one should enter into this phase with the intent of "breaking" the mixture so as to know how far it can be pushed.

The first step in robustness testing is to determine the relationship between fluidity and stability for the materials and proportions chosen. The objective of this testing is to pinpoint at what level the mixture becomes unstable. The following outline provides suggested steps for evaluating this relationship. The slump flow ranges outlined can be adjusted at the practitioner's discretion. Because the relationship between slump flow and other test data is not always linear, a minimum of three slump flow levels is recommended. Each mixture outlined uses identical proportions while adjusting HRWR dosage.

- Mixtures:
  1. Adjust HRWR dosage to achieve 500-550 mm slump flow.
  2. Adjust HRWR dosage to achieve 600-650 mm slump flow.
  3. Adjust HRWR dosage to achieve 700-750 mm slump flow.
- Tests:
  o Slump flow[6]
  o T50 time[6]
  o Column segregation[11]
  o Bleeding[10]
  o Visual stability index (VSI)[6]
  o Air content[13]
  o J-ring passing ability[7]—this test is only necessary when passing ability is a required characteristic.

Upon completion of this testing a number of scatter plots showing the relationship between slump flow, T50, and other test data can be developed for future use, such as the one shown in Figure 9.5. Each data point can be marked with the test result from any of the other tests such as column segregation percentage, VSI, J-ring flow, or others. This figure plots fictitious data for T50 time versus slump flow and column segregation percentage for

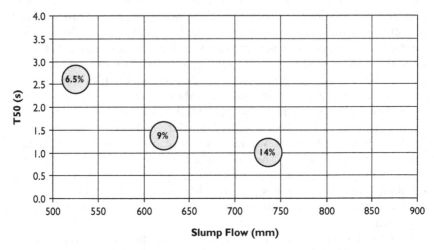

Figure 9.5 Data Plot for Robustness Testing.

the above series of mixtures. In this example, the practitioner can establish slump flow–T50 relationships and quality control limits based on a specified segregation limit. If the segregation limit is 15% or 10%, different parameters will be established for the other two tests.

Another aspect of robustness is a mixture's response to batching errors. This is evaluated by purposefully making changes to the batch weights of a set mixture and measuring the response. ASTM C 94, Standard Specification for Ready-Mixed Concrete, can be used to establish the appropriate changes to the batch weights.[14] The standard provides guidance on material batching accuracy for ready-mixed concrete plants. For example, Section 9 of ASTM C 94 provides the following guidelines for batching accuracy of normal size batches: powders (cement plus mineral admixtures) ±1%, aggregates ± 2%, and total water ±3%. Fluctuating powder and/or water contents can have significant impact on the fluidity and stability characteristics of the mixture. With the intent of ensuring that the breaking point of a mixture is found, it is therefore recommended that the robustness be evaluated at points outside of these accuracy ranges. At a minimum, the author would recommend a further reduction to the minimum powder point (–2% rather than –1%) and an increase to the maximum water point (+6% rather than + 3%). Test results based on these expanded ranges will more fully define the robustness per-formance space, and a greater understanding of the mixture's limits will be attained. Table 9.1 presents a set of example proportions and the minimum and maximum limits of each material.

Control of aggregate-free moisture is often cited as one of the critical areas to control during production of SCC. Unaccounted-for water can dramati-

Table 9.1 Example Material Minimum and Maximum Limits for Robustness Testing

	Example Proportions (kg/m³)	Minimum	Maximum	Minimum (kg)	Maximum (kg)
Cement + Mineral Admixtures	450	–2%	+1%	441	455
Coarse Aggregate	850	–2%	+2%	833	867
Fine Aggregate	800	–2%	+2%	784	816
Water	185	–3%	+6%	179	196

cally affect all of the fresh SCC properties. Therefore, having a clear understanding of a mixture's water robustness is important for the concrete producer. Table 9.2 shows an example robustness testing plan to evaluate a reduction in powder and both a reduction and increase in water. The adjusted values are presented in the shaded areas. In this program, all of the standard fresh property tests are performed and a determination made as to the robustness of the mixture. All admixture dosages are fixed based on the reference mixture, except for HRWR, which is adjusted to achieve the targeted slump flow. A mixture is robust if the batch weight changes result in acceptably small changes in performance as measured by the test methods chosen.

The above methodology, however, will only show the influence of one factor at a time. It should be recognized that real-world concrete production is more dynamic than this. During production of a single batch of concrete, it is possible that the weighed amount of more than one material could fluctuate either higher or lower than the target proportions while still being within the batching tolerances specified. It is possible to evaluate and model

Table 9.2 Example Testing Program for Evaluating Robustness Based on Batching Accuracy

	Example Proportions (kg/m³)	Powder Robustness (–1%)	Powder Robustness (–2%)	Water Robustness (–3%)	Water Robustness (+3%)	Water Robustness (+6%)
Cement + Mineral Admixtures	450	446	441	450	450	450
Coarse Aggregate	850	850	850	850	850	850
Fine Aggregate	800	800	800	800	800	800
Water	185	185	185	179	191	196

these more complicated scenarios; to do so, however, requires a more sophisticated experimental design and set of analysis tools. In a previously published study, this type of robustness analysis was conducted where multiple materials were varied simultaneously.[15] With the materials and mixture proportions used, fluctuations in sand and coarse aggregate weights had minimal impact on SCC performance. The biggest impact occurred when water and total powder weights varied. Therefore, these should be monitored most closely. Additionally, if these factors change simultaneously, different performance fluctuations can occur. For example, if water is batched under target and the total powder is batched over target, the mixture viscosity can be significantly increased. On the other hand, if water is over-batched and total powder is under batched a higher potential for mixture instability and segregation is created. The magnitude of these batch weight variations will also depend on where the initial mixture proportion targets are set. If the target water content is very low to begin with, then in the first scenario mentioned above the viscosity shift may be very large.

Generally robustness is primarily concerned with fluidity, segregation, and bleeding criteria. However, in some circumstances, it may be appropriate to evaluate passing ability robustness by increasing the coarse aggregate batch weight to the maximum value as shown in Table 9.1. At a minimum, however, the powder and water limits should be tested. Testing the aggregate limits is useful, but is left to the practitioner's discretion.

If it is found that the robustness of a mixture is insufficient, one can:

- Reduce the slump flow target level, if possible.
- Increase the paste volume.
- Add or increase the dosage of a VMA. Numerous published reports have concluded that the addition of VMAs to SCC mixtures will enhance robustness.[15-18] This was also shown in Chapter 6.
- Reduce the maximum aggregate size.[17]
- Increase the paste density.[17]

Once a suitable mixture has been developed, the next step is to produce the mixture through the concrete plant and, if possible, place the concrete using the methods and forms proposed for the project.

Production Trial and Mock-Up

Once an SCC mixture has been developed to meet the targeted fresh and hardened properties, a test-pour or mock-up trial is recommended. Although no longer in a laboratory, this step is still part of the development process, just with a broadened scope of testing. One is now evaluating the mixture from the production, delivery, and placement perspective. Focus is still on the SCC mixture properties, but now the following concepts must be considered:

- Batching and mixing on a production scale:
  - o Raw material stockpiling and batching
  - o Moisture control
  - o Mixing concrete:
    - – Mixer type and size
    - – Material batching sequence
    - – Mixing time
- Delivery:
  - o Delivery equipment
  - o Delivery time
- Placement:
  - o Equipment
  - o Techniques
  - o Placement direction
  - o Placement rate
  - o Time
  - o Finishing time (if applicable)
- Personnel:
  - o Training
  - o Batching and mixing
  - o Placement.

In addition to testing one's ability to produce, deliver, and/or place the concrete, the actual benefits of SCC versus conventional concrete can be evaluated at this time. The time required for batching, mixing, placement and finishing can all be measured, as can the human resource requirements for placement and finishing. Other benefits, such as improved aesthetics, resulting in less patching labor and material, can be determined. The author recommends that the producer and/or contractor perform this analysis of benefits during the mock-up phase and then re-visit it during the actual production phase to determine if and by how much any efficiencies have improved.

Careful thought and consideration should be given to the mock-up process. A pre-trial meeting with the quality control, production, and placement crews should be organized. Each function should have a specific evaluative task to perform, and input from all parties should be considered. It is not enough to go from laboratory testing directly to a production environment and surprise the placement crew with a new mixture. If a placement crew is not informed and provided with some minimal training in casting with SCC, how can one determine if an unacceptable surface finish is due to a poor mixture, poor placement technique, or both? The practitioner must be in a position to clearly separate variables in order to answer this type of question.

Summary

The qualification of an SCC mixture includes laboratory testing of fresh and hardened properties, robustness testing, and production mock-up trials or test pours. This process is the link between theory and practice, and it is here that a practitioner can learn a great deal about SCC mixtures and their use. The organizational time and resource investment increases substantially as one moves into mock-ups and test pours, because now the concrete volume and the number of people involved are increasing. As in most things, the success of the qualification process is dependent upon the time investment and preparation prior to execution.

Chapter 10

Concrete Batching and Mixing

Introduction

Producing SCC efficiently and consistently on an industrial scale, while minimizing human involvement in monitoring performance and adjusting proportions, is the next step in the evolution of SCC technology. The transition from mixing SCC in small controlled batches to production in industrial equipment is a significant step with multiple variables that can affect SCC properties. A careful and thorough recognition of these variables will assist in ensuring success when real production and delivery begins.

Industrial-Scale Batching and Mixing of SCC

More consistent SCC will result when good stockpiling, storing, and material batching practices are in effect. A sufficient and consistent supply of the raw materials needed to complete a project is essential. Changing a raw material supplier mid-project may lead to inconsistent performance of the mixture. For example, changing the fly ash source can result in changes to the entrained air content (or air-entraining admixture dosage requirement) and the fluidity of a given SCC mixture.[1] Storage of raw materials should follow industry recommendations, such as those listed in ACI 304 or other equivalent and applicable guidelines. One should ensure that all powder materials are kept dry and free-flowing to permit efficient and accurate batching. A consistent aggregate particle size distribution is important for producing consistent SCC properties. Aggregates should be stored in such a way as to prevent segregation and to ensure consistent moisture content.[2,3] Finer particles in aggregate will tend to settle, particularly during transport, and adequate care should therefore be taken during unloading of aggregates, and aggregate stockpiles should be built up in layers and not end-dumped. One method for limiting aggregate segregation is to separate it into more controllable size ranges with a maximum to minimum size range ratio of less than 4 (for aggregates 25 mm and smaller).[2] This technique also permits greater flexibility in adjusting the aggregate blend to maintain a consistent gradation. It requires, however, a greater number of bins and hoppers for aggregate storage and weighing, and therefore may not be practically realistic for all locations.

Aggregate bins and powder silos should be kept full as much as possible. Filling aggregate bins the night before will allow aggregates to reach a more consistent moisture condition. One caution, however, is in order: if overly wet aggregates are loaded into bins the night before, excess moisture will drain down and increase the moisture content of the aggregates in the bottom of the bin. This may cause fluctuations in the performance of the following day's first few SCC mixtures. In some cases, it is recommended that the first portion of aggregate be removed in the morning if the moisture levels are extreme.

Aggregate Moisture Control

One of the most common concerns during field production of SCC is adequate control of aggregate-free moisture.[4-8] Uncontrolled, excess moisture will increase the water/powder ratio by volume (wv/pv), decrease the mixture's plastic viscosity, and potentially lead to excess fluidity, instability, segregation, and bleeding. Fine aggregate moisture has been found to be stable at levels as high as 6–8%,[2] which could amount to 50–70 kg/m^3 of water delivered through fine aggregate for a mixture using 850 kg/m^3 of sand. It is therefore critical to have an accurate day-to-day and within-day knowledge of moisture contents.

Various methods and equipment are currently used for the control of moisture contents and can be categorized as either manual or automatic. Manual methods include standard test methods, such as ASTM C 566, Test Method for Total Evaporable Moisture Content of Aggregate by Drying, and ASTM C 70, Standard Test Method for Surface Moisture in Fine Aggregate.[9,10] These methods require a technician's effort to test and calculate the moisture content of the aggregates. They can be used when producing SCC as long as significant vigilance is used in monitoring each batch of SCC. The difficulty is that, with these methods, moisture is measured at a single point in time and is assumed to be relatively constant between tests. If potential exists for the moisture content to vary, such as during a rain storm or very dry weather, a rigorous testing protocol, which would include more frequent moisture and concrete testing, should be in place to ensure consistent performance. Some producers store aggregates in covered locations so as to minimize the impact of weather changes. Figure 10.1 shows covered aggregate storage with a misting system for maintaining consistent moisture and Figure 10.2 shows an unprotected aggregate stockpile and the application of water through a sprinkler system. Both producers are attempting to maintain more consistent moisture, but the uncovered system will have a higher degree of variability.

Figure 10.1 Protected Aggregate Storage.

Figure 10.2 Unprotected Aggregate Storage.

Automatic moisture determination methods include the use of moisture probes in aggregate bins, as well as in the mixing vessel. Moisture probes in aggregate bins are generally configured in such a way as to provide feedback to the batching system, which automatically adjusts added batch water to compensate for the aggregate moisture. According to one study, moisture probes have been shown to be reasonably accurate for fine aggregate but not as accurate for coarse aggregate.[5] The Interim Guidelines from PCI require that moisture probes be capable of measuring changes of 0.5% in the moisture content of both coarse and fine aggregates.[11] ACI 304 states that moisture probes should be recalibrated to oven-dried samples monthly or whenever the slump of the concrete produced is inconsistent. Studies have suggested that this calibration procedure should be done with great care, in particular when obtaining the sample for oven-drying. The sample should be obtained from an area as close to the probe as possible.[5] The use of accurate, well-calibrated aggregate moisture probes provides real-time adjustments during the batching and mixing process, leading to more uniform production. In a recent quality benchmarking survey, the National Ready-Mixed Concrete Association in the United States asked producers how often they measure aggregate moisture and how often they calibrate their moisture probes. Eighty-nine percent of the respondents said that they either check the moisture at least daily or use moisture probes. When it came to calibrating the moisture probes, 29% said they did it weekly, 29% said they did it monthly, and 21% said they did it quarterly, while 14% said they did it annually or greater.[12]

Moisture probes can also be installed in the concrete mixing vessel. These probes determine the moisture content of the concrete mixture once all materials have been batched, allowing for trim water addition at the end of the cycle. They can be placed in the mixer either in a fixed position or as a revolving probe attached to the mixing arms. One published study on mixer moisture probes included an experiment incorporating two different water content measurements. The results showed that the probe location (fixed versus revolving) provided different levels of repeatability and total error in water content prediction, with the revolving probe being more precise.[5] During batching, the measured total water content will vary with material addition, as well as homogenization of the mixture. Accurate moisture measurements with in-line mixer moisture meters can take up to 30–45 seconds after all material has been batched for the reading to stabilize.[13] It is recommended that the producer discuss this stabilization time with the moisture meter company's technical representative. Not allowing time for this stabilization to occur may lead to inconsistency in the fresh SCC properties.

Mixing SCC

The batching sequence and mixing process impact the consistency of production of SCC.[14] In addition to the batching sequence, MNL 116 states that the required mixing time for concrete will depend on many factors, including batch size, workability of the batch, size and grading of aggregate, type of mixer, condition of mixing blades, and the mixing efficiency of the mixer itself.[3] Mixing time is the time required to reach a homogenous mixture after all materials are added to the mixer.[2]

The sequence and speed by which materials are added to the mixer can influence the required mixing time.[5,14] Because SCC mixtures have relatively higher fine material contents, sequencing of the powders is important. If fine materials are added first to a drum mixer, head packing may occur. Head packing is the situation where a densely packed agglomeration of fine particles becomes stuck in the head of the drum and does not thoroughly mix into the concrete, resulting in inconsistent properties. To avoid head packing in a drum mixer, ACI 304 recommends adding 10% of the coarse aggregate and water up front. If fine powders are added with water before aggregate addition, balling may occur, and these cement/powder balls may not break down and be thoroughly mixed. The batching sequence will differ based on the mixer and on the production efficiency requirements of the plant. In some instances, a blending of dry materials occurs before the addition of any water; however, the following is a typical batching sequence that has been used to minimize balling in higher powder, low water mixtures in a drum mixer:

1. Coarse aggregates and >50% of mix water
2. Cement and other powders added slowly
3. Fine aggregate
4. Trim water

Admixture addition timing can influence SCC properties such as slump flow level, workability retention, and air content generation. One should follow the admixture manufacturer's instructions for the admixture being used.

SCC can be mixed in all mixer types; however, both the slump flow and the compressive strength can be negatively impacted by insufficient mixing of self-consolidating concrete.[5] Because of the differences in mixing action, the mixing time required to achieve consistent SCC properties will differ by mixer. ACI 304 outlines the following mixer types:

- Drum mixers (Figure 10.3). The typical drum configuration has internal fins attached to the inside of the drum. The mixing action is a folding action of the concrete onto itself. These mixers can either be part of a central mixer or mounted on a truck. Mixing speed is slow compared with other mixer types.[15] These are the least efficient types of mixers, meaning that mixing time to achieve homogenous SCC properties will be longer. *If the SCC is to be mixed or delivered by drum mixer, all wash water must be removed from the drum before batching materials. Control of wash water in truck drums is a key component to consistent*

Figure 10.3 Drum Mixer.

SCC production. The producer should also be aware of any differences between trucks, such as worn fins, which can influence the mixing action and therefore the required mixing time. Using trucks with different mixing efficiency for SCC delivery to the same project may cause inconsistencies in the measured properties. It is recommended that a producer choose and use a controlled set of trucks for a given SCC project.

- Vertical shaft mixers. These can be either planetary mixers with multiple rotating mixing arms attached to a centrally rotating shaft or rotating pan mixers. These types of mixers are frequently used in precast factories and shear the concrete rather than fold it. They mix at medium to high speed and are more efficient at mixing SCC than drum mixers and pugmill mixers. Figures 10.4 and 10.5 show the inside mixing arms of a planetary and a pan mixer, respectively.
- Pugmill mixers. These mixers have a stationary vessel with a horizontal shaft (Figure 10.6). Attached to the shaft are mixing blades that can have different configurations, such as ribbon/spiral blades or paddles. Mixing speed is slow and may require mixing times similar to drum mixers.[16]

Another mixer type not specifically highlighted in ACI 304 is the twin-shaft mixer (Figure 10.7). These are very efficient, high-speed mixers used

Figure 10.4 Planetary Mixer.

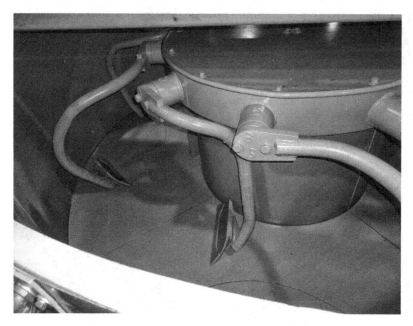

Figure 10.5 Pan Mixer.

Figure 10.6 Pugmill Mixer.

Figure 10.7 Twin-Shaft Mixer.

primarily in precast factories. Mixing time compared with the other types of mixers will be lower. They mix SCC very well.

Volumetric mixers are self-contained units that carry aggregates, cement, water, and admixtures on a single vehicle (Figure 10.8). They are generally used for smaller-volume projects. The mixing occurs in an auger- or screw-type mixer and SCC has been successfully produced in these mixers (Figure 10.9). One should thoroughly test the mixture beforehand, however, since the mixing time is not necessarily adjustable.

The ratio of batch size to mixer capacity will affect the SCC fresh properties and/or the mixing time required to achieve the desired properties. This effect will be more pronounced for less efficient mixers such as drum mixers and as the batch size increases to full capacity. If different batch sizes of the same SCC mixture are run, one should make note of these differences and adjust the mixing process accordingly. Most concrete mixer manufacturers will have a rated capacity for their equipment, and their recommendations should be followed.

The type of SCC mixture being batched influences the mixing time required. Mixtures with higher plastic viscosity tend to require a longer mixing time. In one published study, data were presented on both high-performance and self-consolidating concrete, showing the stabilization time (mixing time required to reach a stable mixer amp reading) as a function of

Figure 10.8 Volumetric Mixer.

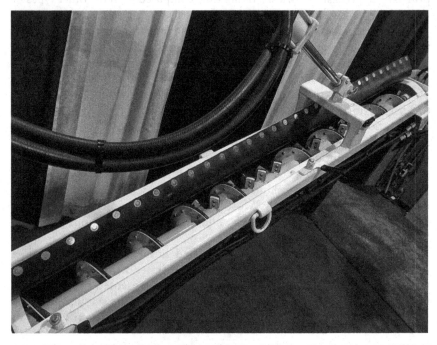

Figure 10.9 Mixing Auger for Volumetric Mixer.

water/cement (w/c) ratio for mixtures with two different cement contents and two high-range water-reducer (HRWR) dosages. The study showed that mixes with a higher w/c ratio required shorter mixing times than lower-w/c mixtures.[13] By using the graph provided in Reference 13, the present author estimated the total water content per cubic meter and plotted it against the stabilization time. Figure 10.10 shows that the total water content of a mixture can be used as a relative indication of mixing time for a given mixer. A concrete producer developing or producing various SCC mixtures would now realize that an SCC mixture with lower water content will require longer mixing to produce consistent properties. All other things being equal, the more water per unit volume of concrete, the more easily mixed it will be.

Releasing the Mixture for Use

Once all of the materials have been batched and mixed, a judgment must be made as to whether the SCC mixture is ready for discharge. Most production facilities with a central mixer, such as precast factories and some ready-mixed concrete plants, monitor the batching and mixing process from a central location such as that shown in Figure 10.11. The batching process, including material sequencing and rate of material addition, is controlled from this

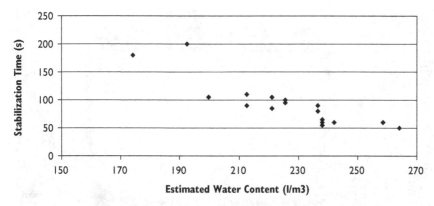

Figure 10.10 Stabilization Time versus Estimated Water Content from Reference 13.

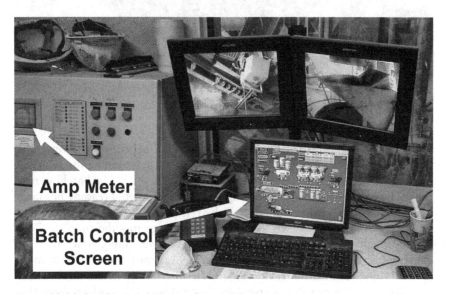

Figure 10.11 Batching and Mixing Control Room.

location. Additionally, once all materials have been added, the batchman in some cases will monitor either the mixer power consumption or the real-time w/c ratio of the mixture in the mixing vessel.

The power consumption of the mixer may be displayed on an amp meter such as the one shown in Figure 10.12, or sometimes it is displayed on the computer screen. The amp meter reading should be given time to stabilize after all materials have been batched. The stabilization time is similar to that

Figure 10.12 Amp Meter Used in the Concrete Batching Process.

of the in-vessel moisture meters discussed earlier. One study has shown that these times may be very similar and can be used to determine the required mixing time for a mixer or set of mixture proportions.[13] The batching personnel will monitor the meter reading, and once a targeted reading has been reached and stabilized, the batch will be released for use. Higher readings indicate that the mixer is working harder to turn or mix the concrete. This is traditionally assumed to result from a stiffer, lower slump concrete.

In other locations pressure gauges, known as "slump meters," are mounted on ready-mixed concrete trucks (Figure 10.13). Slump meters monitor the hydraulic pressure required to turn the mixer drum. The readings from this equipment can be influenced by a number of variables aside from the concrete properties, such as batch size,[13] mixing speed,[17] and presumably mixer blade conditions.

Whether using amp meters or pressure gauges, the same basic principle is being applied: one is using the power consumption or pressure reading as an indication of the concrete mixture's resistance to mixing. This concept is similar to the way in which concrete rheometers function. Previously published studies have revealed the functional similarity of concrete rheometers and slump meters, while others have proposed equations with which to

Figure 10.13 Slump Meter.

predict rheological constants from pan-mixer energy consumption.[13,17,18] The intention here is not to predict rheology from mixing equipment, but rather to more effectively understand and use this equipment to control the production of SCC. To better understand what the data from this equipment mean, a brief review of concrete rheological measurements is necessary.

The rheological constants of yield stress and plastic viscosity were previously discussed in Chapter 3. To determine those parameters for a concrete mixture, a concrete rheometer is necessary, one of which is the IBB Rheometer (Figure 10.14). This functions by measuring a torque value as an H-shaped impeller spins in a planetary motion through a sample of concrete at certain specified speeds. The output of this test is a scatter plot of torque, on the y-axis, versus speed, on the x-axis, through which a line can be drawn. This line provides the rheological fingerprint of the mixture. The y-intercept of the line relates to the yield stress of the mixture, and the slope of the line relates to the plastic viscosity. Figure 10.15 shows fictitious example lines for three concrete mixtures. Lines A, B, and C represent concrete mixtures with different relative rheological characteristics. Mixtures A and B have the same plastic viscosity, indicated by equivalent slopes; however, Mixture A has a higher yield stress (y-intercept) than Mixture B and therefore will have a lower slump or slump

Figure 10.14 IBB Rheometer.

flow. Mixtures B and C have the same *y*-intercept and therefore the same yield stress, and so their slump or slump flow will be similar. Mixture B, however, has a higher plastic viscosity, as can be seen by the greater slope of line B versus line C. It will then have a higher T50 or V-funnel time.

Most central mixers can only mix, and therefore measure power consumption, at a single speed. Therefore, a scatter plot and resulting line (at this point in time) cannot be created. However, for those concrete mixer trucks equipped with a slump meter, the speed can be varied and a scatter plot of

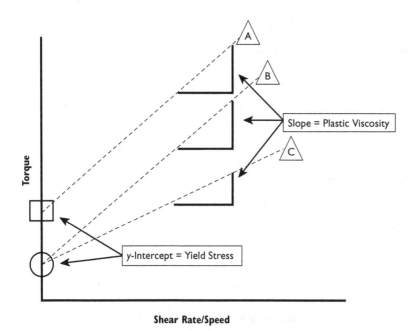

Torque

Shear Rate/Speed

Slope = Plastic Viscosity

y-Intercept = Yield Stress

Figure 10.15 Example Lines from a Concrete Rheometer Data Scatter Plot.

mixing speed versus hydraulic pressure developed.[17] Most producers, however, use the single speed/pressure reading to monitor slump or slump flow, whether in a central batch or on a truck. In doing this, the producer must remember that the slump meter or amp meter reading is influenced by *both* the yield stress (slump or slump flow) and the plastic viscosity (stickiness) of the mixture. Figure 10.16 represents the situation where Mixtures A, B, and C are tested and the torque or power consumption is measured at a single speed. From this single speed point, a vertical dotted line is drawn to intersect lines A, B, and C. This point of intersection correlates to a single torque, amp, or pressure value on the y-axis. This is what happens when a single-point power consumption reading from a concrete mixer is used. In this example, the reading would correctly predict the slump or slump flow difference between Mixtures A and B, because their viscosities are similar. However, because of their plastic viscosity difference, it would not correctly predict the slump or slump flow relationship for Mixtures B and C. In this case, the torque, amp, or pressure reading is higher for Mixture B than for Mixture C, but, theoretically, these two mixtures should have similar slump flows, as indicated by their identical y-intercept/yield stress values. The difference in the single-point measurement is due to the higher plastic viscosity of Mixture B. Essentially, what this tells us is that the technique of using a single speed/power consumption reading can provide an indication of a yield

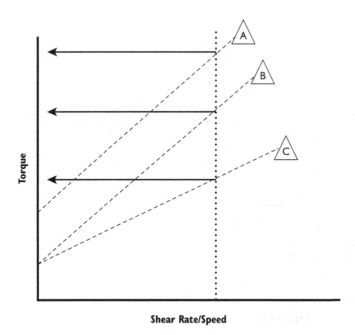

Figure 10.16 Examples of Different Single-Point Measurements for Mixtures.

stress change, if plastic viscosity is known to be constant, or a plastic viscosity change, if yield stress is known to be constant, but not both simultaneously.

The author was involved in a situation where a concrete producer was using an amp meter in this fashion, the results of which are conceptually demonstrated in Figure 10.17. In this facility, certain admixture adjustments were being made without changing mixture proportions. These changes caused an increase in the plastic viscosity of the mixture, while a similar slump was maintained. The batchman was not educated on the rheological changes occurring in the mixture prior to the changes being implemented into the production process. His protocol called for a specific reading on the amp meter before releasing the batch. With the increased plastic viscosity of the mixture (Mixture 2 versus Mixture 1), his amp meter reading was now higher than anticipated. Because the HRWR admixture had already been added, he assumed that the moistures were incorrect and decided to add water to lower the amp meter reading.

When water is added to a concrete mixture, both the yield stress and the plastic viscosity are reduced.[19,20] As water was added to reduce the amp meter reading to the level of the original Mixture 1, a much lower yield stress (Mixture 3) resulted. This was confirmed at the jobsite, as batches were arriving with slumps above the maximum tolerated range. Additionally, compressive strengths were lower than anticipated owing to the added water.

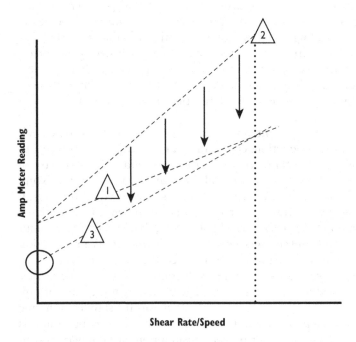

Shear Rate/Speed

Figure 10.17 Example of Misunderstanding Amp Meter Output.

Therefore, if changes are made to a concrete mixture, the possible effects on the batching and mixing process must be anticipated and communicated. This can be significant with SCC, particularly if different types of SCC mixture are being produced or if viscosity-modifying admixtures (VMAs) or other admixtures that can influence plastic viscosity are included. The amp meter or slump meter reading should be calibrated to the slump flow and T50 time of each SCC mixture to be produced.

Improving the Approach

When it comes to consistently producing SCC, or any concrete mixture, one area in need of improvement is with respect to the control of the batching and mixing process. Improvements can come in the form of batch plant operator training, which will be discussed in Chapter 14, as well as improvements to equipment and monitoring practices. The key to establishing an effective industrial batching process is to choose the appropriate leading indicators of performance and quality. These leading indicators are monitored while the batch is still in the mixer and can be adjusted if necessary. Sophisticated industries, such as the chemical industry, have clearly established targets and monitor dozens of variables during production to ensure final production quality. For example, during the production of certain polymers, as raw materials are fed into a reactor, the producer may monitor temperature, mixture pH, re-circulating pump pressures, and other variables to ensure that the chemical reaction is proceeding at the correct rate and achieving the correct parameters. In concrete production, however, many producers do not have a leading indicator of product quality in the mixer, those that do use either the amp meter/slump meter or the in-vessel moisture meter, and the typical producer uses only one of these.

In addition to monitoring the mixer power consumption and batch w/c ratio during concrete production, conceptually, it would also be possible to monitor the reading stabilization time and use it as an indication of the batch quality. As noted in Figure 10.10, the stabilization time for a given mixture (of certain batch size and mixer type and size) is influenced by the overall water content of that mixture: the more water per batch, the shorter the reading stabilization time. Therefore, theoretically, those producers without moisture meters inside of their mixing vessels could use this as a production monitoring technique to generally confirm the correct water content.

Monitoring multiple variables during production is one option for further improving SCC production consistency. The author is not aware of any concrete production software that specifically monitors the stabilization time and uses it as a production control; therefore, the two measurements at this time would have to be the mixer power consumption and the w/c ratio of the batch. If a concrete production plant can automatically monitor both of these, theoretically, they can be used together for greater production control. A matrix can be created where the power consumption reading is plotted on the x-axis and the w/c ratio reading is plotted on the y-axis (Figure 10.18). The center points, represented by the solid lines, of both axes are the target readings for the mixture being produced and the chain-dotted lines are the acceptable variation as determined by practice and experience. This matrix assumes a set of mixture proportions that have been produced with some regularity and that no significant raw material changes have occurred.

The lower left quadrant in Figure 10.18 is defined as an area where the power consumption is lower than targeted, with water lower than targeted. This indicates that the mix is low in water but easier to mix. As has been discussed, all things being equal, lower water will reduce the slump flow level and increase the plastic viscosity of a mixture, typically resulting in a higher power consumption reading. Therefore, in this case, quality assurance/quality control actions are required to investigate and determine the cause of this discrepancy. Similarly, the upper right quadrant represents an area where the moisture or w/c ratio is reading higher than normal but the power consumption is also reading high. All things being equal, higher water generally leads to higher fluidity and lower plastic viscosity, which would result in a lower power consumption reading. Matrix readings in these two quadrants indicate more serious production problems that should be investigated.

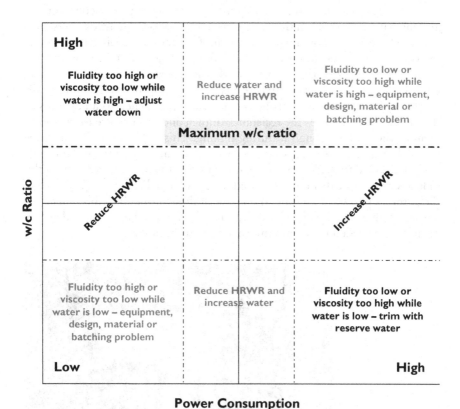

Figure 10.18 Automated Moisture or w/c Ratio Reading and Power Consumption Matrix.

Matrix readings in the upper left or lower right quadrants in Figure 10.18 represent areas where water adjustments can be made. In the upper left, the moisture reading is too high and the power consumption low, indicating that water should be reduced on subsequent batches. In the lower right, total mixture moisture is low and power consumption is high, meaning that the trim water should be added. In the case where the moisture reading is on target but the power consumption reading is higher or lower, the HRWR dosage can be adjusted to increase or decrease the slump flow and move toward the center of the matrix. However, if the power consumption reading is on target but the moisture readings are higher or lower than targeted, one can make the initial adjustment to both water and HRWR to move toward the center of the matrix; however, results in these areas can also represent a potential material, equipment, or batching issue that should be investigated. This relationship, if used, should be established for each SCC mixture and plant location. At the present time, the author is aware of at least one concrete batch control manufacturer that can monitor both the power consumption and the batch w/c ratio on the same screen, but not on a matrix as presented above. The author is not aware of any manufacturer that specifically monitors the stabilization time and uses it as a leading indicator of batch quality.

Other batch monitoring equipment that is reported to measure the rheological characteristics of the mixture has been developed and is commercially available.[21] It consists of a probe (Figure 10.19), attached to an arm inside the mixer, that measures the resistance of the concrete mixture to the movement of the probe through it during mixing. This information is converted to rheological data using specialized software. Currently, this probe is only being marketed to precast producers, because of the mixer types used in this industry versus the drum mixers used in ready-mixed concrete. Although to the author's knowledge a limited number of these probes have been installed in precast plants in North America, this is still the type of equipment development that is needed to further improve SCC technology.

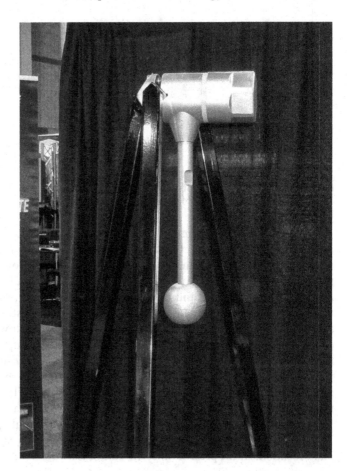

Figure 10.19
Probe Arm for
Attachment into
a Concrete
Mixer.

Other Considerations

Very viscous mixtures tend to require a longer mixing time[5,13] and can place a relatively higher stress on the mixer. ACI 304 states that the mixer should be designed for starting and stopping under full load conditions, and the majority of mixers can do this. One report, however, indicated that mixtures with very low wv/pv have tripped breakers during production and shut down mixers. In this case, reducing the batch size will reduce the power draw of the mixer and keep this from happening.[1]

Faster mixing does not guarantee better mixing or a shorter mixing time. An example of this is in the area of high-performance concrete (HPC). High-performance concrete is used on bridge decks in the United States and is often proportioned using silica fume and a low w/c ratio. For these applications, the concrete is delivered to the jobsite in mixer trucks. These mixtures can be quite viscous and, during mixing, may adhere to the drum mixer walls. In a guide for HPC, published by the Federal Highway Administration, it is recommended that for truck-mixed HPC, because of the "sticky" nature of the mixtures, the drum mixing speed be slowed and the batch size reduced to 70% of the capacity to allow for more efficient mixing.[22] The lower mixing speed will allow the slow-flowing, viscous mixture to fold and mix more efficiently and not simply stick to the drum wall and spin. This same concept can apply to high versus low plastic viscosity SCC mixtures.

Summary

The transition from small, controlled laboratory experiments to production of SCC using industrial equipment is a significant step in the development of any SCC program. A well-thought-through transition process should consider everything from storage and stockpiling of materials, batching equipment, and accuracy, to the mixing equipment, and finally to the protocol for releasing a batch for use. It is impossible to provide a step-by-step guide for mixing every type of SCC mixture using every type of storage, batching, and mixing equipment. The practitioner should therefore strive to understand the nuances of his or her own equipment and facility, and how the particular SCC mixture to be produced may respond during the production process. The purpose of a carefully controlled transition is to ultimately be able to produce SCC of good, consistent quality efficiently and with minimal human intervention, either during the process or afterwards in troubleshooting field problems. A consistent and repeatable batching and mixing process is a key component to producing consistent-quality SCC. Deviations from the process should be recorded and communicated in the event that any troubleshooting of mixture performance needs to occur. A clear communication process should be established as a part of the quality control program.

Chapter 11

Quality Control of SCC Materials and Mixtures

Introduction

A well-constructed quality control plan is as important to an SCC program as establishing the correct performance targets or developing an optimal set of mixture proportions. Given time, the producer or contractor that views quality control as an invaluable resource and not just a necessity will likely find it easier to produce consistent SCC. This is because they will put the data generated to use and learn from it. ACI terminology defines quality control as "the actions taken by a producer or contractor to provide control over what is being done and what is being provided so that the applicable standards of good practice for the work are followed."[1] With SCC, good quality control is about more than just testing and generating numbers, it is about turning data into information and putting that information to use in refining and improving the SCC program. This is especially true in the early stages of SCC production and use, when experience levels are low. The quality control effort required at these stages will likely be higher than that required for conventional concrete. A heightened level of attention must be paid to one's materials, the mixture's fresh properties, and their impact on the concrete's final in-place quality. How much additional quality control attention is needed for SCC is somewhat dependent upon the performance requirements of the mixture. Similar to high-strength or other high-performance concrete mixtures, the level of attention given increases as the performance requirements increase and the allowable margin for error decreases. For example, as an SCC mixture's slump flow increases, so does its potential for segregation. SCC mixtures with higher slump flow targets will require relatively higher levels of attention to ensure the mixture stability when compared with a similar mixture with a lower slump flow requirement. As time progresses and experience with SCC increases, the producer develops an understanding of the interrelationship between materials and the performance of mixtures. Once this level of expertise has been reached, quality control maintains the same importance but now becomes more intelligent regarding control of the variables that are most critical for a specific application. This is the level of expertise one should aspire to.

In establishing the outline for a quality control plan around SCC, one should realize that a plan is most effective when developed and agreed upon before the start of a project, and when it is well communicated among the entire team. Quality control is the responsibility of all members of the construction team. The suppliers of aggregates, admixtures, cements, and other powders are responsible for ensuring that their materials meet the needs and requirements of the project and the concrete producer. In turn, the concrete producer is responsible for the quality and consistency of the batched and delivered concrete. Finally, once the concrete is on site and in use, it is the contractor's responsibility to ensure the quality of the in-place material. When everyone is clear on their responsibilities from the beginning, the project has a much higher likelihood of success. These general concepts are no different for SCC than they are for conventional concrete, and for all those involved a well-conceived quality control plan may result in some of the following benefits:

- A highly successful project and quality end product
- Development of a higher level of internal material and SCC performance expertise
- Elimination or reduction of post-batching or post-placement trouble-shooting
- Creation of more economical SCC mix designs or casting methods
- Increased market and customer use of SCC
- Further development of a reputation as a technically astute concrete professional.

The suggestions in this chapter are meant to enhance, not supersede, any existing quality control program. The intent is to outline areas specific to SCC that should be monitored before and during production and which then become additions to the currently existing program. Areas covered include monitoring of raw materials such as aggregates and powders, reviewing case studies and published guidelines on monitoring of concrete performance, and establishing acceptance criteria. Finally, an internal auditing process is suggested for the producer.

It is worth repeating here that the purpose of quality control testing is to ensure the integrity, performance, and service life of the final structure, but the quality control process also provides an opportunity to further understand material influences on SCC performance. In particular, if a batch falls outside of the targeted performance range, the producer would be well served to invest time to discover the reason why. Although this process requires time and resources, it is the author's opinion that understanding why something did not work is many times as important as understanding why a thing does work. It is also through these efforts that an idea for a new piece of production monitoring equipment or some other new development may occur. If permitted, failures and mistakes do represent opportunities.

Raw Material Quality Control

In general, any data collected during the quality control process, whether it is from raw material or concrete testing, serves two purposes: the first is to provide an immediate indication that the material's or concrete's quality is ensured; the second is to create a link between concrete performance and material characteristics. This is a longer-term goal as performance and testing history is developed. Analysis of historical data becomes useful in the selection of raw materials and proportions for future projects and may also highlight the need for mixture proportion adjustments.

The powders and aggregates used to produce SCC should be monitored regularly. Monitoring these raw materials can be done by collecting data from the material supplier and/or result from an internal testing program. The tests performed and data collected should be the same as those collected during the raw material characterization activities completed during the laboratory mixture development phase. In this way, if fluctuations are experienced in the performance of an SCC mixture, one has the opportunity to correlate raw material changes to the changes in concrete performance.

In the results of a recent quality benchmarking survey published by the National Ready-Mixed Concrete Association in the United States, survey recipients were asked a number of questions about their quality control program and practices.[2] This survey was not specific to SCC but to all concrete production. When asked at what frequency are aggregate grading tests run per plant the responses were that 25% use the aggregate supplier's data, 54% test it weekly, 14% test monthly, and 7% test quarterly or less frequently. When asked if information was collected on the primary cement used, 4% responded that no data were collected, 64% said they collect the cement company mill report data, 25% said they collect the mill report plus other data, and 7% said they simply use other cement company information. When asked about the frequency with which they test the cementitious materials from their primary sources, 57% said they do not test, 4% said they test weekly, 7% said they test monthly, and 32% said they test quarterly or less frequently. From conversations with some precast and ready-mixed producers, it appears to the author that these statistics are close to the norm for SCC production as well.

A thorough outline of procedures and practices for monitoring the consistency of aggregates, cement, and other raw materials used in the production of concrete are outlined in numerous industry documents.[3,4] The following sections, however, highlight those characteristics of raw materials that should receive consideration when SCC is being produced. If not already in place, these considerations can then be included into one's overall quality control program.

Monitoring Powders

Some powders used in SCC may be more variable than others. For example, if a powder used originates from a manufacturing waste stream, such as crusher fines from aggregate production, versus a more controlled production process, such as Portland cement, the more variable material should be monitored more closely and tested more often. Proactively testing powders, however, is not an activity that most concrete producers engage in, as can be seen from the survey results described above. Accurately characterizing powders requires some specialized, accurate equipment, as well as human resources and time. Typically, a producer will monitor powder influence through concrete mixture performance. Data generated in this way will not always provide clear guidance regarding powder performance, depending on the thoroughness of the concrete testing. If a producer does not have the ability to test powder characteristics, he or she should request reports on the physical and chemical characteristics of the powders from the respective suppliers. The producer should also ensure, as much as possible, that the data provided relate to the lot of material currently in use. Collecting data to establish the historical record of a material's characteristics is also important. As an example, Table 11.1 shows Blaine fineness data for several lots of cement from three suppliers.

As one collects material data in conjunction with concrete data, certain historical trends may emerge. For example, the use of cement with a more variable fineness, such as Cement A in Table 11.1, could lead to less consistent fresh properties, fluctuating water requirement, and/or high-range water-reducer (HRWR) dosage. This historical data can also assist in the selection of which material to use on future projects based on their consistency. Or, if the option of choosing alternate materials does not exist, then adjustments may be made to the mixture proportions, such as increased use of alternate powders or increases in viscosity-modifying admixture (VMA) dosage, to enhance robustness and reduce the required quality control effort. Common quality control tools such as histograms of material characteristics are good methods for monitoring consistency.

Table 11.1 Statistical Analysis of Blaine Fineness Data for Three ASTM C 150 Type I Portland Cements

	Cement A	Cement B	Cement C
Mean (m²/kg)	391	351	385
Standard Deviation (m²/kg)	54	17	23
Range (m²/kg)	161	47	67
Minimum (m²/kg)	300	321	335
Maximum (m²/kg)	461	368	402
Coefficient of Variation	14%	5%	6%
Count	11	7	8

Monitoring Aggregates

If the properties of aggregates change dramatically, changes to fresh SCC properties will likely occur. It is assumed that as long as the production process and aggregate composition do not change, aggregates from the same source will have a reasonably consistent particle shape. The main aggregate properties to be monitored for SCC include gradation or particle size distribution and void content, and, if the absorption is typically high, monitoring changes to the absorption characteristics may be warranted. In addition to these static properties, the aggregate-free moisture content should be monitored and controlled.

Monitoring Particle Size Distribution

Sieve analysis to determine aggregate gradation should be, and typically is, routinely conducted on the fine and coarse aggregates used for SCC production. The quality control manual for the National Precast Concrete Association requires that gradation tests be conducted for approximately every 1360 metric tons of fine aggregate and 1800 metric tons of coarse aggregate. But how can aggregate gradation information be used to enhance the quality and consistency of SCC? Monitoring the percent passing the 300 and 75 µm sieves is suggested and further described below.

In Chapter 5, it was shown that the total material in a mixture passing the 300 µm sieve (from aggregates and powders) can be used to gauge the bleeding potential of an SCC mixture. Although the bleeding tendency of SCC is dependent on mixture proportion, monitoring this characteristic helps one to predict if and when a mixture adjustment, such as VMA addition or dosage increases, are required. For example, Figure 11.1 plots the percent passing the 300 µm sieve of over 100 separate lots of fine aggregate from a single source.

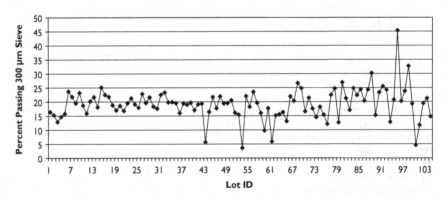

Figure 11.1 Percent Passing the 300 µm Sieve for Fine Aggregate 1.

Notice that in some cases, a significant drop or increase from one lot to the next occurs. For example, Lots 52 and 53 have 15% and 4% passing, respectively. This is obviously a significant change, and because it is such a radical shift, the test on Lot 53 should probably be repeated to ensure its validity before making any mixture proportion changes. Let us assume, however, that it is a valid result. How would this impact the performance of an SCC mixture? Let's now assume one has developed an SCC mixture with a fine aggregate content of 850 kg/m³. If one is producing SCC with Lot 52 and then shifts to Lot 53, this would result in a 100 kg/m³ reduction in material passing the 300 μm sieve. Depending on how the original mixture was proportioned, this variability could result in a significant change in the bleeding tendency, as well as the robustness of a mixture batched with identical proportions (see Figure 5.3 as an example).

The variability of this characteristic will differ by material source; therefore, collecting and analyzing historical data on this property for multiple sources is a valuable exercise. Table 11.2 shows a comparative analysis of the percent passing the 300 μm sieve for two fine aggregates. Fine Aggregate 2 has a lower coefficient of variation than Fine Aggregate 1. However, even with a relatively less variable fine aggregate, one may still experience fluctuations in performance. This is where monitoring and review of historical data becomes valuable from a mixture proportioning perspective. Although Fine Aggregate 1 is more variable, it also provides a higher percentage of fine material than Fine Aggregate 2, meaning that the amount of fines added from another source may be reduced. This can influence the total material cost of a mixture. A producer will need to make a decision here between material cost and the ability to monitor material and adjust proportions as necessary. If one does not have this historical data during the initial proportioning phase, then, as this monitoring occurs, mixture adjustments can be made and performance improved. Since the variability is greater, Fine Aggregate 1 should be tested more often than Fine Aggregate 2 until such a time that its consistency improves.

Table 11.2 Comparative Analysis of Percent Passing the 300 μm Sieve for Two Fine Aggregates

	Fine Aggregate 1	Fine Aggregate 2
Mean	19.20	14.10
Standard Deviation	5.40	2.10
Range	41.7	8
Minimum	3.6	10.2
Maximum	45.3	18.2
Count	104	25
Coefficient of Variation	28%	15%

The material passing the 75 μm sieve is considered to be part of the powder and therefore paste content of an SCC mixture. Ensuring the consistency in powder and paste volume is critical to a consistently performing SCC mixture. Figure 11.2 plots the percent passing the 75 μm sieve for 90 lots of Fine Aggregate 1. Significant shifts between successive lots are also seen for this characteristic. For example, Lots 50 and 51 have 3.2% and 0.3% passing the 75 μm sieve, respectively. This is a difference of 2.9%. Again assuming an SCC mixture with 850 kg/m³, as we move from Lot 50 to 51, a reduction of approximately 25 kg/m³ of powder occurs. With a fine aggregate specific gravity of 2.6, this equates to almost a 1% reduction in paste volume. Depending upon the characteristics of the lot of fine aggregate used to proportion the original mixture, one could experience an increase in mixture viscosity, water, or admixture demand and a reduction in flow if the amount of fine material increases. Conversely, if the fine material decreases, one may experience a decrease in stability, water or admixture demand or an increase in flow.

Collecting and reviewing historical data on this characteristic is also valuable. Table 11.3 provides a comparative analysis of the material passing the 75 μm sieve as determined by ASTM C 117 wet sieving for the same two fine aggregates.[5] The variability of the two aggregates is similar based on the coefficients of variation; however, because Fine Aggregate 2 contributes very little of this material overall, the magnitude of its impact is lower than that of Fine Aggregate 1. During the mixture proportioning phase, if one includes this fine material from the fine aggregate in the powder content, then knowing its variability is critical. Steps should be taken to proportion a mixture with sufficient amounts of other powders so as to ensure that the paste volume does not fall below a certain minimum level. Regardless, in this case again, Fine Aggregate 1 should be tested more often owing to the magnitude of the potential changes.

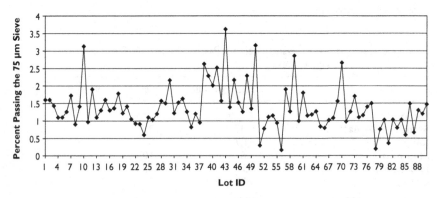

Figure 11.2 Percent Passing the 75 μm Sieve for Fine Aggregate 1.

Table 11.3 Percent Passing the 75 μm Sieve for Two Fine Aggregates

	Fine Aggregate 1	Fine Aggregate 2
Mean	1.37	0.47
Standard Deviation	0.63	0.23
Range	3.45	0.95
Minimum	0.17	0.11
Maximum	3.62	1.06
Count	90	23
Coefficient of Variation	46%	48%

In the majority of cases where a quality control plan exists, aggregate gradations are regularly conducted. The author suggests that histograms of the percent passing the 300 and 75 μm sieves be created and monitored. Although currently no predictive equations linking performance of all SCC mixtures to these measurements exist, relative relationships do, as described above, and, as a producer develops greater experience with SCC, he or she will begin to know when a certain increase or decrease in these values will cause performance problems.

Where passing ability is a strong concern, the producer should monitor the percent retained on the larger coarse aggregate sieves. A significant increase in the quantity of these larger size fractions could result in a greater blocking potential and a decrease in passing ability. The critical levels can be calculated using the equations provided by Billberg.[6]

Monitoring of the Aggregate Void Content

In SCC, the paste volume is the primary driver of the fluidity and flowing properties of the mixture. The aggregate void content will influence the volume of paste required. An increase in the void content will in effect reduce the amount of free paste available to promote flow. One can monitor the void content of the coarse and fine aggregates independently or one can monitor the void content of the aggregate combination that is being used to produce a given SCC mixture. The latter option is more specific to SCC and likely most useful. All other things being equal, changes to the combined aggregate void content will be more directly relatable to concrete performance. An increase or decrease in void content will likely result in a decrease or increase in mixture flow, respectively. The frequency of testing will be dependent upon the in-place quality control plan or it can be determined by knowing the variability of the raw material source.

Monitoring of Free Surface Moisture

Most modern production facilities will continuously monitor moisture contents, either in the aggregate bins or in the mixing vessel (see Chapter 10). Consistent and accurate moisture control is critical for consistent SCC production. The quality control manual for the National Precast Concrete Association states that surface moisture content shall be physically tested once a day prior to the first SCC batch, even when moisture probes are in use. If in-line moisture meters are not used, both PCI and NPCA require that the moisture be manually measured at the beginning of each batching operation and every 4 hours of continuous batching or at any time a change in moisture content becomes apparent. A very robust mixture should be developed for production facilities that do not use in-line moisture meters.

Monitoring Concrete Performance

To ensure that a concrete testing plan is as efficient and effective as possible, it should be agreed upon in writing before the start of a project. The plan should outline the specific testing required, those responsible, and the frequency of testing.[7] Some or all of the SCC properties of fluidity, passing ability, stability, and viscosity may be measured for production control or site acceptance purposes, and certain considerations should be given when testing these properties as part of a quality program.

Fluidity

Ensuring that a minimum level of fluidity is achieved and maintained from production to placement is essential. The slump flow test is a standard measure of this property and is typically, if not always, included in quality control plans for SCC. The minimum and maximum limits, as well as the acceptable range for slump flow values, depend on the nature of the project. In previous chapters, the relationship between slump flow and both stability and surface finish were discussed. In some cases, the maximum slump flow must be closely controlled so as to ensure the intended stability of the SCC mixture. This will be more important the higher the target slump flow is. For example, if a project has a slump flow target range of 500–600 mm while another has a target range of 700–800 mm, the higher-performance mixtures will be more susceptible to fluctuations and should be monitored more closely. If a high-quality surface finish is required, one must ensure that the slump flow does not fall below an established minimum value. Lower slump flows than those targeted could result in voids being trapped on the as-cast surface. The maximum and minimum values for slump flow should be established during the mixture proportioning and qualification stages. An acceptable range of values is one that is not overly restrictive, but still ensures that a mixture will produce the intended results. For the three consistency classes outlined in the European SCC Guidelines, the slump flow ranges are 180, 160, and 160 mm for classes SF 1, 2, and 3 respectively. This means that if one were to target the median value of these ranges the targets would be ± 90, 80, and 80 mm, respectively.

Passing Ability

Passing ability testing should only be part of a quality control plan when the project demands it. A variety of passing ability test methods have been developed, including the U-box,[8] the L-box,[7] and the J-ring.[9] All of these tests can be performed in the field. The author has seen early specifications for SCC that listed multiple passing ability tests in the required test method section. It was clear that at the time of writing the specification, the specifier was not familiar with the details of the test methods and so listed several. This was overly burdensome for both the producer and contractor. It is these situations that the pre-project discussions on quality control will hopefully avoid.

In addition to these tests, an apparatus configured with multiple obstacles representing reinforcement was developed in Japan for continuous testing of SCC passing ability.[10] The apparatus is placed on top of the pump hopper and, as SCC is discharged from the ready-mixed truck, it must pass through this apparatus. The benefit of this method is the continuous testing of all SCC being placed.

Stability

Stability testing in the field or on the production site is limited to those methods that can be performed with reasonable speed and not overly delay the placement of concrete. For that reason, ASTM has indicated that the column segregation test[11] is to be used for laboratory testing and not field control; however, ASTM C 1712[12] can be used in the field. In addition to the penetration apparatus test, the visual stability index (VSI) should be included as part of any quality control plan. Although this test is limited, it is easy to perform and requires no additional equipment or time. If a trained technician is determining the VSI, he or she can provide very useful information on the mixture performance, particularly if problems occur. The original intent of the VSI was for exactly this purpose. It was not developed as a means of characterizing or quantifying stability, even though a number is given based on the visual rating. This method was a result of witnessing hundreds of SCC mixtures and clearly recognizing (like most SCC practitioners) that a visual examination of a mixture does provide very real information and to ignore it is not wise. The VSI is a valuable method for monitoring the consistency of the mixture's fresh properties. If the VSI changes during production, it is clear evidence that something else is changing, especially when the slump flow stays constant.

Viscosity

Changes to the viscosity of a mixture can be monitored by using the T50 test[13] or other tests such as the V-funnel or O-funnel. Of all these tests, T50 is the least accurate, but requires the least amount of additional labor, time, and equipment at the casting site. The T50 test is a useful tool for monitoring the consistency of produced and delivered concrete.

Case Study Analysis

Numerous tests can be conducted as part of a quality control plan, but how does one determine which tests should be conducted on a given project? In reviewing numerous case studies, some describe the on-site acceptance testing program used. Table 11.4 shows the results of this review by highlighting the properties measured and the methods used. From this sample of case studies, we see that the fluidity and viscosity are the two most widely measured characteristics for job site control and conformity testing. Stability was rarely measured on site via a quantitative test method. One reason for this is that, until recently, there has not been a reliable and rapid method for quantifying the stability of a mixture in the field. This has changed somewhat with the introduction of the penetration apparatus test.[12] One can now quantify the stability of an SCC mixture with reasonable effort and accuracy. Prior to this, the VSI was and still is one of the primary means of *qualifying* a mixture's stability during production. Additionally, as has already been mentioned, there is a relationship of both slump flow and viscosity with stability, and some rely on this relationship to ensure stability.

In addition to the case studies, most published SCC guidelines provide direction regarding concrete testing at either the production plant or the casting site. The following highlights some of the requirements from a selection of industry-accepted guidelines.

- *Recommendations for Self-Compacting Concrete*, Japan Society of Civil Engineers.[28] The JSCE recommends that the slump flow, T50 or V-Funnel, and U-box tests be conducted at the plant for production control and on-site before placement.
- *The European Guidelines for Self-Compacting Concrete*. These guidelines suggest that the producer test slump flow at the production site on each batch of SCC until consistent results are obtained and that other tests may need to be conducted (such as for passing ability) as the project demands. Visual inspection of each batch is also recommended prior to delivery, even when consistent slump flows have been documented.

Table 11.4 Published Site Acceptance Testing and Methods from Case Studies

Ref.	Fluidity	Passing Ability	Stability	Viscosity	Frequency	Other
14	Slump flow					
14	Slump flow					
14	Slump flow					
14	Slump flow					
15		Testing apparatus		O-funnel	Every 70 m^3	Amp meter used to control production
16	Slump flow			T50		Amp meter used to control production
17	Slump flow			O-funnel	Every 70 m^3	Amp meter used to control production
18	Slump flow			T50		Amp meter used to control production
19	Slump flow			T50		
19	Slump flow			T50		
20	Slump flow			T50		Amp meter used to control production
21	Slump flow					
22	Slump flow					
23	Slump flow	Caisson test		V-funnel		Concrete rheometer also used
24	Slump flow		Sieve and/or column segregation			
25	Slump flow	L-box		T50		
26	Slump flow	Kajima test		V-funnel		
27	Slump flow	L-box or J-ring	VSI	V-funnel		One of either L-box, J-ring or V-funnel also required
27	Slump flow	L-box or J-ring	VSI	V-funnel		One of either L-box, J-ring or V-funnel also required
27	Slump flow	L-box or J-ring	VSI	V-funnel		One of either L-box, J-ring or V-funnel also required
27	Slump flow	L-box or J-ring	VSI	V-funnel		One of either L-box, J-ring or V-funnel also required

On-site, the slump flow test and a visual assessment for any indication of segregation are recommended.

- *Quality Control Manual for Precast and Prestressed Concrete Plants*, National Precast Concrete Association (NPCA). These guidelines recommend both the slump flow and VSI tests for production and acceptance testing. The slump flow and VSI tests are to be performed each day on the first batch of SCC, and on consecutive batches until two consecutively produced batches are within specification. Thereafter, slump flow and VSI testing shall be performed every 50 yards or 25 batches, whichever comes first, or when changing mix designs or raw materials or when a mixture becomes suspect or a problem occurs. More frequent testing is required for plants not using automated aggregate moisture monitoring equipment.

- *Interim Guidelines for the Use of Self-Consolidating Concrete in Precast/ Prestressed Concrete Institute Member Plants*, Precast/Prestressed Concrete Institute (PCI).[29] PCI recommends that the slump flow and VSI tests be conducted for production and acceptance testing. The T50 test is an optional test, suggested to be run twice per month. The frequency of the slump flow and VSI testing is the same as that required for slump testing in the PCI Quality Control Manual.[3] This includes testing at the start of the production day when making compressive strength specimens, whenever the consistency of the concrete seems to vary, and at least one time for every three air content tests.

Further Suggestions

The guidelines and case studies reviewed recommend or report using the slump flow as part of both the production and job site acceptance testing. Some measurement of the concrete viscosity and a visual examination of the fresh concrete for signs of instability are also typically recommended or reported. It is the author's opinion that if the slump flow is being tested, then the slump flow, T50, and VSI should be reported together. The beauty of this is that all three measurements are made while performing a single slump flow test. If a more accurate measure of the viscosity is needed, then the O- or V-funnel can be used. Similarly, if more accurate data on stability are required, then the penetration apparatus can be used.

When these tests are performed together, a matrix of the slump flow, T50, and VSI data can be created. Figure 11.3 presents an example matrix, which includes an outlined target area between 650–750 mm slump flow and 1–2.5 s T50 time. This provides the quality control technician with an immediate visual confirmation that the mixture's performance is good, as long as the stability is acceptable. To include stability, we see that four example test results are plotted and the numbers for each point correspond to the VSI of the mixture. The VSI becomes an important indicator in this scenario to assist

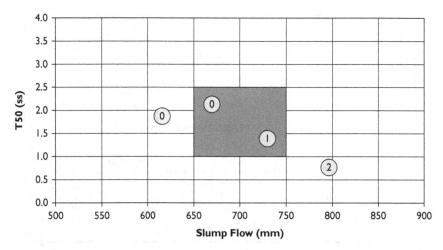

Figure 11.3 Quality Control Matrix of Slump Flow, T50 and VSI Testing.

in the troubleshooting process. By using this matrix, a producer begins to see a clearer property relationship between fluidity, viscosity, and stability.

As a producer establishes this relationship and collects data over time, the information can be used to create a chart for making mixture adjustments. An example chart is shown in Figure 11.4. In this example, the background shows the likelihood of instability and segregation for the particular mixture being produced and tested. This chart does not relate to every SCC mixture. It is a producer's responsibility to create these tools based on their materials and mixtures. For the situation under consideration, one would expect, for the mixture that has been developed, characterized, and used in production, that if a tested batch results in a slump flow and T50 relationship falling into the lightly shaded area, a low likelihood of segregation exists. The medium shaded area, however, indicates that some potential for segregation exists and the heavily shaded area indicates that there is a strong likelihood of segregation.

Based on where a set of test results fall relative to the target area, further guidance can be provided for adjusting subsequent batches to achieve the slump flow and T50 target. That guidance is outlined in some detail surrounding the target area within the chart. Relative to the target area, the areas above, below, above right, and below left represent areas where multiple variables may be interacting, causing the results to miss the target, and more detailed troubleshooting may be necessary. The areas to the left, right, above left, and below right represent areas where some potentially straightforward adjustments can be made. It is here that the VSI ratings become useful as confirmation that the mixture adjustments made are reasonable.

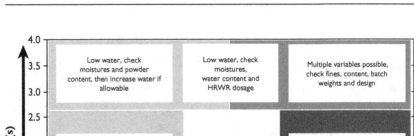

Figure 11.4 Example Slump Flow and T50 Matrix with Mixture Adjustment Guidelines.

For example, if a batch in the above-left area, which should be stable, is also showing signs of instability, then the adjustments outlined in this matrix will not be correct and a more detailed investigation into the batch weights, moisture contents, or other variables is needed.

Any producer can build these types of charts and tools for the SCC mixtures they are producing if the data are consistently collected and reviewed. When it comes to testing production consistency, the slump flow, T50 time, and VSI together can provide a reasonable indication of any significant changes to the concrete performance. Other tests, such as those for passing ability or for more accurate confirmation of stability or mixture viscosity, can be added to the quality program as determined by project requirements.

Linking Quality Control Testing to Production

With some effort, a producer can link the information from concrete quality control testing to the production acceptance matrix presented in Chapter 10. In this matrix, the power consumption reading from the concrete mixer and the water/cement (w/c) ratio as determined by in-vessel moisture probes are plotted on the x- and y-axes, respectively. Again, in this quality control scenario, we are assuming that a single set of mixture proportions is being used, no radical adjustments or additions of admixtures are occurring, and the material properties are reasonably constant. If this is the case, it is reasonable to assume that the T50 time (as an indication of viscosity) will be heavily influenced by the water/powder (w/p) ratio. As the w/p ratio increases or decreases, the T50 time (viscosity) will decrease or increase, respectively. The power draw from the mixer has been successfully used to estimate the slump flow.[30,31] Because we are also measuring and predicting viscosity changes, we can assume that as the power draw on the mixer increases or decreases, a decrease or increase in slump flow is occurring, respectively. Linking a T50 target range to a w/p ratio or water content (depending on the specific batching and mixing system), and then the slump flow target range to the mixer power draw, will, over time, provide more clear and specific targets for the production and mixing personnel.

Data points will be in opposite positions on the production acceptance and quality control testing matrices: Figure 11.5 shows the two matrices and the correct relationship between areas on each. For example, area A on the production testing matrix represents a mixture with a high T50 and low slump flow, and on the production matrix this would likely result in a batch with a lower w/c ratio and higher mixer power consumption. This type of relationship would need to be developed for each mixture a producer creates. The producer would also need to be aware that the batch size produced will influence the mixer power consumption. If quality control testing results can be effectively correlated to production equipment readings, then, over time, the concrete producer will become more efficient in his or her concrete production and will require less quality control time to be invested in batch adjustments and production monitoring.

Internally Auditing the Consistency of SCC Production

Once SCC is in production, it is a valuable exercise for the concrete producer to schedule a testing program where the quality control department randomly chooses one or two days over which to test all of the SCC properties: fluidity, stability, passing ability, and viscosity for a given mixture. This is essentially equivalent to an audit of the production and quality control process. The author has participated in several of these programs with various producers, and this subsection provides examples of the levels of consistency that can be experienced when producing and using SCC. Three examples of SCC production in three separate precast plants (Companies A, B, and C) are shown. The testing programs were conducted over a one- or two-day period during which a team of technicians were present to randomly select samples of production SCC. Prior to these auditing programs, no inputs on the mixture proportions or performance targets were given; however, after the audit, recommendations for improvement were provided to each producer as necessary. The tests performed included slump flow, T50, VSI, V-funnel, and column segregation.

Company A produces precast wall panels for the residential market. Panels are cast horizontally with numerous box-outs for windows and doorways. Table 11.5 shows the data collected during production. The performance

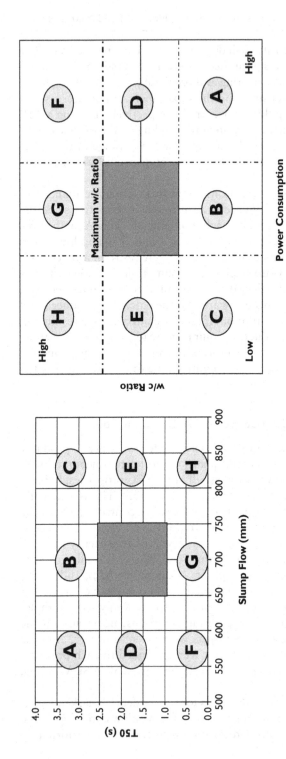

Figure 11.5 Relating the Production and Quality Control Testing Matrices.

Table 11.5 Company A Production Data

Batch	1	2	3	4	5	6	7	8
Slump Flow (mm)	595	585	595	570	560	585	NA	NA
VSI	0	0	0	0	0	0	NA	NA
T50 (s)	5.01	5.43	4.88	5.22	5.63	5.42	NA	NA
J-Ring (mm)	535	535	545	525	515	540	NA	NA
J-Ring Difference (mm)	65	50	50	45	45	45	NA	NA
V-Funnel (s)	7.02	7.22	5.22	9.47	6.63	6.97	NA	NA
Column Segregation (%)	NA	NA	NA	NA	NA	NA	2.8	1.9

NA: not available.

targets for Company A were originally described by slump flow only, with a target of 550–650 mm. During the testing program at Company A, it was not possible to perform the column segregation test on the same samples that were tested for all of the other properties. Therefore, no analysis of this data will occur, except to say that the two batches tested showed very good stability.

Table 11.6 provides a statistical analysis of production consistency based on the set of data presented. It should be noted that the results of this analysis, and those that follow, are influenced by both production and testing variability. The V-funnel results, in this example, are not as consistent as one would like. However, based on the consistency of the slump flow and T50 test results, the V-funnel variability may be influenced more by test error than production error. The slump flow and T50 measurements were very consistent. Combining these results with the separate column segregation and VSI results, the overall production control and quality of SCC at Company A were considered to be very good.

Company B produces wall panels for commercial buildings. A slump flow of 550–650 mm was the only SCC performance target set for Company B's production. The first batch in Table 11.7 shows that the slump flow was below the target range. An adjustment was made to the following batch that increased slump flow above the desired range, decreased the T50 and V-funnel times, and increased the column segregation percentage and VSI

Table 11.6 Statistical Analysis of Company A's SCC Production Data

	V-Funnel (s)	*Slump Flow (mm)*	*J-Ring Difference (mm)*	*T50 (s)*
Mean	7.09	582	50	5.27
Standard Deviation	1.37	14	8	0.28
Range	4.25	35	20	0.75
Minimum	5.22	560	45	4.88
Maximum	9.47	595	65	5.63
Coefficient of Variation	19%	2%	15%	5%

Table 11.7 Company B Production Data

Batch	1	2	3	4	5	6	7
Slump Flow (mm)	500	660	595	615	635	635	650
VSI	0	2	0	0	0	0.5	1
T50 (s)	NA	0.9	1.3	1.2	1	1.2	1.1
J-Ring (mm)	475	635	565	580	595	610	625
J-Ring Difference (mm)	25	25	30	40	40	25	25
V-Funnel (s)	1.97	1.67	2.12	2.02	1.87	1.82	1.86
Column Segregation (%)	7	10.7	4.2	5.3	7.9	9.7	8.6

NA, not available.

rating. A second adjustment was made, after which the slump flow was more consistent and fell within the targeted range. Toward the end of the placement, the slump flow began to drift higher and the T50 and V-funnel values drifted lower, causing the mixtures to lose some of their stability, as can be seen in both the column segregation test as well as the VSI.

Table 11.8 shows a statistical analysis of the SCC production from Company B. Overall the analysis of Company B's production indicated a reasonable level of control. Although the coefficient of variation of the column segregation testing was high, the variability occurred within a reasonably acceptable range of values. It was recommended to Company B, however, that they investigate the process by which they determine when and how to make batch-to-batch adjustments during production. Particular attention should be paid to production of the day's initial batches.

Company C is a prestressed double-tee producer and the only performance target for their production is a slump flow of 500–600 mm. At this facility, the concrete is delivered from the batch plant to the casting beds via mixer trucks. Table 11.9 presents the production data collected. The first, third, and fourth batches did not achieve the minimum slump flow target out of the mixing plant. It was noted that water was added to these three batches to increase the slump flow after the fresh property testing samples were taken

Table 11.8 Statistical Analysis of Company B's SCC Production Data

	Column Segregation (%)	V-Funnel (s)	Slump Flow (mm)	J-Ring Difference (mm)	T50 (s)
Mean	7.63	1.9	613	30	1.12
Standard Deviation	2.32	0.15	54	7	0.15
Range	6.5	0.45	160	15	0.40
Minimum	4.2	1.67	500	25	0.90
Maximum	10.7	2.12	660	40	1.30
Coefficient of Variation	30%	8%	9%	24%	13%

Table 11.9 Company C Production Data

Batch	1	2	3	4	5	6	7
Slump Flow (mm)	380	510	470	440	545	585	570
VSI	NA	0	NA	NA	0	0	0
T50 (s)	NA	3.23	NA	NA	3.11	2.87	3.05
J-Ring (mm)	NA	430	NA	NA	470	535	525
J-Ring Difference (mm)	NA	75	NA	NA	75	50	45
V-Funnel (s)	NA	4.68	NA	NA	4.42	3.58	4.11
Column Segregation (%)	NA	5.1	NA	NA	9.4	11.3	10.6

NA, not available.

and before placement. This is not a recommended practice for SCC. If any adjustments to slump flow were necessary, it was recommended they be made with additional HRWR and that they be tested again so that accurate records could be kept.

Table 11.10 shows a statistical analysis of the data collected. Because of the three batches that did not achieve the required minimum performance, the number of test results for several of the test methods is limited to four samples. Overall, the consistency of production at Company C was poor and required improvement. A recommendation was given to further assess the overall quality control program, including material controls, at this facility. A recommendation to halt the procedure of re-tempering the SCC mixtures with water was also given. It was also recommended that each plant include the T50 and VSI as targets for production. It is the author's opinion that slump flow by itself is not enough.

A quality control testing matrix for Companies B and C was created from the data collected and is presented in Figure 11.6. Since the column segregation data were available, they were used rather than the VSI for this matrix, and a maximum 10% limit was used for these examples. Company B's results are identified by the shaded squares. Based on these results, to ensure mixture stability, they should target 575–650 mm and 1–2 s for slump flow and

Table 11.10 Statistical Analysis of Company C Production Data

	Column Segregation (%)	V-Funnel (s)	Slump Flow (mm)	J-Ring Difference (mm)	T50 (s)
Mean	9.1	4.2	500	61	3.07
Standard Deviation	2.76	0.47	74	16	0.15
Range	6.13	1.1	205	30	0.36
Minimum	5.13	3.58	380	45	2.87
Maximum	11.26	4.68	585	75	3.23
Coefficient of Variation	30%	11%	15%	26%	5%

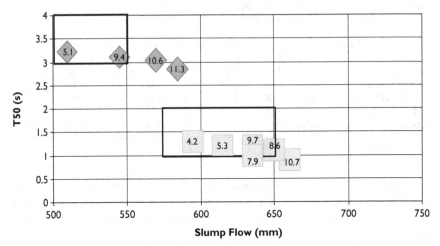

Figure 11.6 Quality Control Matrices for Companies B and C.

T 50, respectively. Company C's results are identified by the shaded dia-
monds. To ensure stability, they should be targeting a 500–550 mm slump
flow and a 3–4 s T50 time. These targets, especially the 50 mm slump flow
range, could be quite restrictive for Company C. This, combined with the
overall audit results, suggests that Company C should immediately invest
some time in improving their SCC program. Mixture adjustments and quality
control procedure developments are likely required.

Summary

The quality control process is intended to be a seamless process from control
of raw materials to acceptance testing at the casting site. The process ensures
that the intended performance of the concrete is present in both its fresh and
hardened states. In addition to any current quality control program, suggestions for quality control of SCC would include the following:

1. Monitoring and control of materials, including fineness of powder
   materials, percent passing the 300 and 75 μm sieves in the fine aggregate,
   void content of the aggregate combination being used, and control of
   moisture content in both fine and coarse aggregates.
2. Production control of the SCC fresh properties. This includes the appropriate calibration and use of production monitoring equipment such as
   mixer amp meters and automated moisture meters (measuring either
   aggregate or concrete batch moistures). As a minimum, the first batch of
   the day should be tested and each batch thereafter until two consecutive
   batches meet the fresh performance targets. Any test data should be

correlated back to the production monitoring equipment in a continuous process to refine the production equipment targets.

3. If automated equipment is not available for monitoring moistures or production, then a plan including additional concrete testing should be instituted. For a ready-mixed concrete producer in this situation, the slump of the mixture before the addition of the HRWR can be measured, and then, after the addition of the HRWR, the slump flow, T50, and VSI (or some other set of tests for monitoring fluidity, viscosity, and stability) should be measured prior to the truck leaving the plant. For a precast producer in this situation, once the mixture has been discharged from the mixer, making adjustments may not be possible. Verification of the concrete in the mixer, although time-consuming, may be valuable in the long run. Any verification of concrete in this way should be done so as to ensure the complete safety of everyone involved. The frequency of the additional concrete testing will likely be more intense initially until a history of production and control has been established. However, as a minimum, the SCC mixture should be tested two to three additional times daily once consistency has been established.

4. If passing ability is a critical performance characteristic for the given project, then monitoring this property is also appropriate.

5. On-site acceptance testing should include those tests that pertain to the project requirements. One should refrain from requiring multiple test methods that monitor the same or a similar property, such as using both the J-ring and L-box tests. It is still appropriate, as a minimum, to measure the slump flow, T50, and VSI. In this way, a direct relationship between the plant and site measurements can be made.

6. An internal production audit program, where fluidity, passing ability, stability, and viscosity testing is conducted on production concrete, will evaluate the health of the current in-place SCC and quality control programs.

Chapter 12

Transport, Placement, Finishing, and Curing

Introduction

The purpose of this chapter is to review the methods and concepts associated with transporting, placing, and curing SCC. Some of the items regarding transport of SCC range from delivery size and logistics for cast-in-place concrete to the use of non-agitating equipment to transport SCC in a precast facility. The placement of SCC is also a significant portion of this chapter, and a review of factors affecting surface finish will be presented. It should be noted that this book does not outline all of the safety factors associated with transporting and placing SCC. All of the standard precautions found in the accepted industry guidelines should be followed, in particular with regard to formwork construction. Additionally, although a variety of placement techniques and concepts will be discussed, this chapter is not intended to limit the techniques used to those presented, nor is there any implied guarantee that, if followed, the placement outcome will be acceptable, as that is determined by many factors.

Transporting SCC

ACI 304R-00 states that the method of transportation (conventional concrete) should be chosen to efficiently deliver the concrete to the point of placement without significantly altering the concrete's desired properties. When selecting the method of transportation, one should consider the mixture proportions and properties, placement method, access and rate of placement, the required delivery capacity (rate and total), the location of the plant relative to the casting site, and weather conditions.[1]

For cast-in-place projects, SCC should be delivered in agitating equipment such as a revolving drum mixer. Keeping the concrete agitated during transport will help maintain homogeneity and reduce the potential for segregation. SCC must arrive at the casting site ready to be placed without the need for consolidation; for this reason, the workability retention of the mixture must be designed to match the delivery and placement time.

The delivery logistics of SCC may be different from that of conventional slump concrete requiring consolidation. The following items should be considered during development of the delivery schedule:

- Total volume of concrete required for placement.
- Placement rate. Typically, the placement rate of SCC will be faster than that of conventionally vibrated concrete.
- Delivery batch size. Because SCC is more fluid than conventional slump concrete, the relative delivery batch size may be reduced to avoid concrete spillage during transport. This is particularly the case in hilly terrain. The concrete mixer drum should be revolving at such a rate during transport so as to limit the potential for spillage to the greatest degree possible.[2]

Given the faster placement rate and the potentially smaller batch size, the concrete producer will need to determine the optimum number of trucks to use for a given project. This will be based on the total placement volume along with the travel time and distance from the batch plant.

On some conventional concrete projects, the high-range water reducer (HRWR) is added to the concrete at the jobsite. This is an option for SCC as well; however, in order to ensure consistent SCC fresh properties, it is recommended that the mixture be batched to SCC consistency at the plant. The only time that HRWR should be added on site is if the slump flow is below the target value. If an SCC delivery arrives with too low a slump flow, water should never be used for increasing it, as this can lead to segregation, bleeding, and other problems.

In a precast facility, concrete can be transported from the mixer to the casting bed in several ways:

- Revolving drum mixer (Figure 12.1). This can usually transport the largest delivery volume of those listed and maintains agitation of the SCC mixture, ensuring greater homogeneity and less potential for segregation.
- Auger discharge transport vehicle (in North America, a trade name for this equipment is Tuckerbilt®; Figure 12.2). This can transport a moderate volume of SCC per delivery. Minimal agitation is possible except for near the auger at the bottom of the hopper. Care should be taken to ensure that the SCC is stable and can withstand the drive from the mixer to the casting beds. This is particularly important if the terrain over which the equipment must drive is not paved and will impart jostling and other energy into the SCC mixture. Once the hopper is filled, engaging the auger in the reverse direction and tilting the auger arm upward will reduce any tendency for paste to separate from the mixture down the auger arm.[2] The top of the hopper is open; therefore, any placement outside during even a light rain must take into consideration the water being added to the top portion of the SCC mixture.

- Bucket and forklift. This has a smaller delivery size than the drum mixer or the auger discharge transport vehicle. This method is also non-agitating and therefore the stability of the mixture must be ensured. This is particularly important if the terrain over which the forklift must drive will impart jostling and other energy into the SCC mixture. It may be necessary to place a gasket between the edges of the "clam shell" opening of the bucket so as to minimize or eliminate paste leakage.[2] Similar to the auger discharge transport vehicle, buckets are open at the top and therefore placement outside during rain must be controlled so as to minimize impact on the concrete performance.
- Bucket and crane (Figure 12.3). Again, this has a smaller delivery size than the drum mixer or the auger discharge transport vehicle. It is also non-agitating, and therefore the stability of the mixture must be ensured. However, the transport of buckets by crane is typically smoother and will impart minimal energy into the SCC mixture, thereby reducing the potential for segregation during transport.

For buckets and other non-agitating equipment, ACI 304 recommends a maximum delivery time of 30–45 minutes.[1] This is for workability retention (slump loss) and homogeneity concerns. With SCC, one should also ensure that the mixture does not display thixotropic properties when non-agitating equipment is used, as this could cause problems during delivery and placement.

Placement

Once the SCC mixture arrives at the point of discharge, it can be placed by any of the means used for placing conventional slump concrete (excluding

Figure 12.1 Revolving Drum Mixer.

Figure 12.2 Auger-Discharge Transport Vehicle.

Figure 12.3 Bucket and Crane.

those for projects without formwork, such as very stiff paving concrete). In many cases, the placement technique will be dictated by the transport equipment used. The following points can be made regarding standard methods of placement:

- Revolving drum mixer. Delivery from a revolving drum mixer can result in further flowing distances and enhanced filling capacity and is effective in placing a wide range of fluidity levels.[2]
- Crane and bucket. Concrete will fall straight down out of the bucket; however, a chute for directing the flow of the concrete can be connected to the bottom of the bucket if the concrete is required to flow horizontally.[2] The same applies to the auger discharge transportation vehicle.
- Pumping. There are numerous case studies in the literature where SCC has been pumped. Pumping of SCC should be done slowly at first in order to limit the potential for paste/aggregate separation in lower-viscosity mixtures.[2,3] For higher-viscosity mixtures, pumping should be started at a lower pressure until concrete flow begins; once the mixture is moving, pumping rates can increase. Another option, which will result in a better surface finish, is to pump SCC from the bottom of the forms.[3] This, however, is not always possible.

Regardless of the method used for placing SCC, there are certain variables that are common to all techniques that should be addressed, including the following:

- Continuous or discontinuous placement? This addresses the question of whether the form will be filled without stopping or whether there will be pauses between successive deliveries. If placement is to be discontinuous, one must know the number of batches to be delivered and the elapsed time from the first to the final placement, as some SCC mixtures can stiffen rapidly owing to loss of workability or to thixotropy. If placement is to be discontinuous, the time-dependent workability properties should match the placement timing. Additionally, a higher slump flow should be used for discontinuous placements (such as those using a bucket) to ensure that the concrete from each delivery flows to its maximum and self-consolidates with the previously placed batches.[2]
- Assess the impact of the relative potential energy of the placement technique. Are the volume and rate of placement relatively high? Higher volume and rate during placement can permit the SCC mixture to flow further in the formwork. Higher-slump-flow mixtures should be used for placement techniques with relatively low placement volumes and rates.
- Placement flow direction versus filling flow direction. As much as possible, the direction of flow from the placement equipment should be in the same direction as one desires the formwork to be filled. Figure 12.4 shows two

Figure 12.4 Placement Flow versus Form-Filling Flow.

pictures of SCC being placed into the stem of a prestressed double-tee. The picture on the left shows the concrete placement flow moving in the same direction as the form-filling flow. The picture on the right shows the placement flow going in the opposite direction to the form-filling flow. Having the placement flow and form-filling flow moving in the same direction can enhance the flow of the concrete down the form and improve surface finish and casting efficiency. If the placement flow and form-filling flow are in opposite directions, turbulence can occur and small air voids may be entrapped within the concrete being placed.

- Placement turbulence. When concrete, or any liquid, is dropped into itself, air is drawn into the in-place material. Figure 12.5 demonstrates this with water being poured into a jar through a small hose. The picture on the left shows the jar before the addition of water. The middle picture shows the turbulence and air being drawn into the liquid as the water is allowed to fall into itself from above. The picture on the right shows the absence of turbulence once the water reaches a height to cover the hose. What one cannot see is that the air bubbles that are created during the turbulent placement can escape owing to the low viscosity of the water. However, a higher-viscosity material such as SCC does not release the entrapped air as easily or as rapidly. As placement continues, these entrapped voids can move to the form surface. This is important for SCC, in particular for those placements where surface finish is critical. To avoid this turbulence, the discharge point should be placed below the surface of the concrete. This may not always be possible; if it is not, then the drop height should be reduced as much as possible and the placement rate should also be reduced. If these last two actions are taken, one must be sure that a high slump flow (700 mm) is being used (which is typically appropriate where blemish-free surfaces are desired). Be aware that this turbulence can also happen as concrete is being dropped from a mixer into a bucket or other transport equipment.

Figure 12.5 Placement Turbulence Drawing Air into Water.

- Placement location. It is preferable to place SCC so that it flows in a single direction throughout the duration of the pour. Pouring should be started and the concrete allowed to flow. Then, as the SCC starts to build in height (if the slump flow level is relatively low), the discharge point should be slowly moved to just in front of the maximum height (Figure 12.6). Note that the location where concrete is being deposited moves as the element is filled. This is the standard technique used during production of most prestressed double-tees. It is also typical (in North

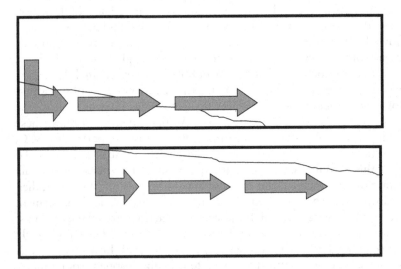

Figure 12.6 Moving Placement Location During Casting.

America) in these applications that the slump flow is between 500 and 600 mm.

- Placement or discharge rate. The importance of the placement rate depends on several factors, including the element dimensions, presence of obstacles, and the flow (rheological) properties of the mixture. Even at a high slump flow, a highly viscous mixture (slow-flowing) should be placed more slowly in order to allow air voids entrapped during discharge to escape; if it is poured too fast, the entrapped air can stay trapped within the concrete and possibly move to the form surface. If the dimensions of the formwork are narrow, then the placement rate should match the rate of flow within the form, or the concrete will build up too high in the form at the discharge point, potentially entrapping more voids. The more narrow the dimensions within formwork, the more influence the mixture viscosity will have on the final results, and a highly viscous SCC in this scenario should be placed more slowly.
- "Box-outs" or other obstacles. A box-out is a formed opening located in the center of formwork and around which the concrete must flow and fill during casting, such as a window or a doorway. Box-outs can occur in walls or panels cast both vertically and horizontally. Those in vertical formwork pose the greater challenge. It is very important to ensure that the concrete flowing underneath a box-out is allowed to flow under in one direction and fill up the other side (Figure 12.7). Do not pour from one side and then move to pour from the other side of a box-out before the underneath is totally filled, or a void may be trapped underneath.
- Flowing distance. How far should one allow the concrete to flow from the discharge point? Less viscous mixtures may have a tendency to separate the further they flow down the form. This should be considered

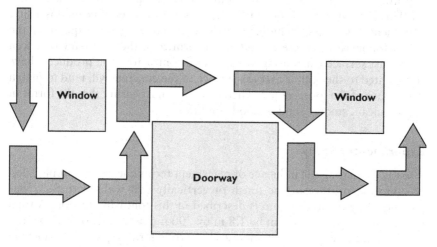

Figure 12.7 Pouring around "Box-Outs."

during the mixture proportioning and qualification stages. In one study, Lange et al. measured the aggregate content of cores taken at various distances from the casting location, and after 13.5 m the coarse aggregate content dropped severely.[4] Another study on the use of SCC in pre-stressed beams demonstrated that when cores were analyzed, the coarse aggregate concentration in the beams was most different between the bottom of the beam at the end where placement started and the top of the beam where placement finished.[5] In this particular situation, one must remember that an SCC mixture in a highly reinforced section will have its passing ability continuously tested until that area of the beam is fully cast. Therefore, if the SCC must flow a long distance through many constant obstacles, the dynamic stability of the mixture is critical. ACI 237 suggests a maximum of 10 m flowing distance and RILEM suggests 5–8 m.[2,3]

- Drop height. When considering drop height, one should consider if the concrete is falling through dense reinforcement or not. The viscosity of the mixture will be important to ensure that the mixture does not segregate during dropping. RILEM references real-world experiences of dropping SCC 8 m, but drops of 1–3 m are more common.[3] Limiting the drop height will also limit the potential for splashing of paste (in both vertical and horizontal forms), which may cause surface blemishes or discoloration.

Vertical As-Cast Surface Finish

One of the many benefits of using SCC is the enhanced surface appearance of formed surfaces. Numerous variables can influence the final surface appearance when using SCC, including form oil type,[6] application of form oil (Has it been applied too thickly?), slump flow level, mixture viscosity, placement techniques (including vibration), and finally surface quality of the forms. If imperfections are present on the formwork, they will tend to be seen in the cast surface when using SCC. Several contractors and producers have commented to the author that the surface of the concrete will tend to match the surface of the forms in use. Therefore, one must ensure that the forms in use are clean, smooth, and in good condition.

Experimental Study

A study evaluating the influence of placement technique and the application of vibration on the surface finish of vertically cast walls with SCC was conducted by the author and is described in this following section. A total of 15 walls measuring 1.8 m by 1.8 m by 200 mm were cast over two days. The wall forms used were common steel/plywood forms typical for residential and commercial walls. They were not coated or lined with special material to produce a glossy finish and no reinforcement was present. The form release used was oil-based. A number of mixture and placement variables were evaluated over the two days of testing, including:

- Mixture proportions.
- Slump flow level.
- Placement techniques:
  o Pumped through a 75 mm line a distance of approximately 15 m— concrete was placed both by free-fall from the top of the form and by immersing the pump line 200–300 mm below the concrete surface.
  o Discharged directly from the truck chute and dropped from the top of the wall form.
- Application of vibration. Two walls were cast with each placement method. Of the two walls, one was vibrated and the other was not. Each vibrated wall was cast in two 0.9 m lifts and an internal vibrator with a 40 mm head was inserted into four equally spaced insertion points dropped to the bottom of the lift and removed for a total insertion time of 12 s per point.

In total, four deliveries of concrete were used to cast the 15 walls. The concrete from the first truck was used to cast Walls 1–4. No change was made to the concrete from Wall 1 to Wall 2, but, before casting Walls 3 and 4, additional HRWR was added to the truck in order to increase the slump flow. Similarly, a second truck delivered the concrete for Walls 5–8. The as-delivered SCC was used for Walls 5 and 6, but, before casting Walls 7 and 8, additional HRWR was added to increase the slump flow. For Walls 9–12, a single delivery of concrete was used without modification to the batch, while for Walls 13–15, a single unmodified delivery was used.

Once cast, and de-molded the following day, each wall panel was divided into nine 610 mm by 610 mm squares (Figure 12.8) for surface appearance evaluation. Each square was given an individual rating according to the following criteria:

0. Glass-like finish with no bugholes
1. Glass-like finish with few small (< 3 mm) bugholes
2. Multiple bugholes, small and medium (3–12 mm) in size
3. Multiple bugholes, some large (>12 mm), medium and small
4. Multiple bugholes, predominantly large
5. Honeycombing, poor consolidation.

These nine ratings were then averaged to give the overall rating for the wall panel itself. Table 12.1 gives the mixture proportions, fresh properties, placement techniques, and surface rating data for the 15 walls.

Top Left	Top Center	Top Right
Center Left	Center Center	Center Right
Bottom Left	Bottom Center	Bottom Right

Figure 12.8 Wall Panel Matrix for Rating Surface Appearance.

The results of this experiment can be considered from several perspectives:

1. Concrete mixture proportion influences. Although the rheological properties were not measured in this evaluation, the mixtures were intentionally proportioned to have varying rheological properties and therefore a different likelihood of producing acceptable surface finishes. This was accomplished by changing paste volumes, water contents, water/powder ratios by volume (wv/pv), and viscosity-modifying admixture (VMA) dosages. Trucks 1 and 2 had the lowest paste volumes, the lowest water contents, and the lowest wv/pv ratios. The difference between these two trucks was that the VMA dosage in Truck 1 was higher than in Truck 2. Truck 3 had slightly higher paste volume, water content, and wv/pv ratio than Trucks 1 and 2 and the same VMA dosage as Truck 2. Truck 4 supplied a mixture with the highest paste volume, water content, and wv/pv ratio and the same VMA dosage as Trucks 2 and 3. Table 12.2 shows the average rating of the walls cast from each truck regardless of placement technique used. The results matched the expected outcome based on the mixture proportioning variables. Mixtures with higher paste volumes and higher wv/pv ratios will typically produce a lower viscosity and have a higher likelihood of producing a good surface finish, assuming similar placement techniques (as long as segregation does not occur).

2. Slump flow level. In previous chapters, we saw that quality surface appearance was tied to higher slump flow values for many SCC case studies. In this evaluation, walls were cast from Trucks 1 and 2 at two slump flow levels and a higher slump flow did not improve the surface

finish in these cases. It is possible that the limiting factor in these tests was the mixture proportions, in particular the lower wv/pv ratio and paste volume. It is also possible that the slump flow will have a greater impact in walls or sections that are more intricate rather than simply flat.

3. Placement technique. During casting of Walls 1–8, 11, 12, and 15, the concrete free-fell into the form regardless of whether it fell from the truck chute or from the pump line. For Walls 9, 10, 13, and 14, the concrete was placed by pump, with the pump line submerged into the concrete until the form was completely filled in order to eliminate free-fall of concrete into concrete. The walls that were placed by free-falling in general had poorer surfaces than the tremied walls. Three direct comparisons between free-fall and tremie can be made (Table 12.3). In each case, the tremie method resulted in an improved surface appearance. Walls 9 and 13 had the best surface finish of the 15 walls cast. We also see that the influence of placement technique was reduced as the mixture proportions were changed to enhance surface finish from Truck 3 to Truck 4. This further highlights the interdependence of mixture properties and placement techniques relative to surface finish.

4. Application of vibration. In this evaluation, there were seven direct comparisons of the influence of vibration on the surface finish of the cast walls. In six of the seven instances, the average surface appearance rating was worse when the SCC was vibrated. It should be noted that the vibration technique was constant across all the walls cast. Therefore, the conclusion here is that vibration using the applied technique tended to worsen the surface appearance. It is possible that, with shorter lifts, more prolonged vibration insertion time, or a larger vibrator, the surface finish could have been improved, especially for those cases where the concrete was allowed to free-fall. One should always be aware of the stability and segregation issues surrounding vibration of SCC and proceed accordingly.

5. Location in forms. All the walls were filled from the center and the concrete was allowed to flow out to either side. On average, the vertical center third of each cast wall tended to have a slightly better surface rating than either the left or the right side. Additionally, the bottom of each cast wall tended to have a slightly improved surface rating. Figure 12.9 shows the average ratings by location on the wall across all 15 walls cast.

The primary conclusions from this testing is that there is an interdependence between placement technique and mixture properties with respect to surface finish. All other things being equal, mixtures with higher wv/pv ratios and higher paste volumes tend to provide better surface finishes. The best surface finish is seen when the concrete is placed by the tremie method, not allowed to free-fall, and not vibrated.

In many cases when casting vertical elements, the pump line cannot be extended to the bottom of the form because of the presence of reinforcement

Table 12.1 Mock-Up Wall Placement Technique and Surface Finish Experiment

Truck	1a	1b	2a	2b	3	4
Type I Cement (kg/m³)	398	399	393	394	395	411
Class F Fly Ash (kg/m³)	70	70	70	70	69	73
Fine Aggregate (kg/m³)	938	939	918	921	931	849
12.5 mm Coarse Aggregate (kg/m³)	760	761	748	750	761	751
Water (kg/m³)	165	166	162	162	172	200
VMA (ml/100 kg)	489	489	391	391	391	391
PCE HRWR (ml/100 kg)	691	822	782	958	1500	587
Slump Flow (mm)	650	735	620	690	590	615
Air (%)	1.4	1.1	2.3	2.4	2.4	2.8
Paste Volume %	34%	33%	34%	34%	35%	39%
wv/pv Ratio	1.0	1.0	1.0	1.0	1.1	1.2

Wall	1	2	3	4	5	6	7	8	9	10	11	12	13	14	15
Placement Technique	Chute	Chute	Chute	Chute	Chute	Chute	Chute	Chute	Pump	Pump	Pump	Pump	Pump	Pump	Pump
Tremied/Dropped	NA	NA	NA	NA	NA	NA	NA	NA	Tremied	Tremied	Dropped	Dropped	Tremied	Tremied	Dropped
Vibration	No	Yes	No	Yes	No	Yes	No	Yes	No	Yes	No	Yes	No	Yes	No
Surface Appearance Rating (0–5)															
Top Left	2	2	2	2	—	3	2	2	—	—	3	3	1	2	—
Top Center	2	—	2	2	2	2	2	2	0	—	0	3	0	—	0
Top Right	2	—	2	3	2	3	2	2	0	—	1	3	—	3	—
Center Left	2	—	2	3	—	3	—	2	2	—	3	2	—	—	2
Center Center	—	—	—	3	2	2	—	2	0	0	2	3	0	—	—
Center Right	2	2	2	2	2	3	2	2	—	2	2	2	—	2	—
Bottom Left	2	2	2	2	—	2	2	—	1	2	2	2	—	—	—
Bottom Center	—	1	2	2	—	2	—	1	0	—	1	2	0	—	—
Bottom Right	1	2	2	2	1	2	1	2	0	2	2	3	1	—	1
Rating Analysis															
Average Rating	1.7	1.4	1.9	2.3	1.4	2.4	1.6	1.8	0.4	1.2	1.8	2.7	0.7	1.4	1.0
Minimum Rating	1	1	1	2	1	2	1	1	0	0	0	2	0	1	0
Maximum Rating	2	2	2	3	2	3	2	2	2	2	3	3	1	3	2

Table 12.2 Average Surface Finish Rating for All Walls by Truck/Mixture

	Truck 1	Truck 2	Truck 3	Truck 4
Average Wall Rating	1.8	1.8	1.5	1.0

Table 12.3 Surface Finish Comparison Between Tremie Method and Free-Fall

	Truck 3				Truck 4	
Wall	9	11	10	12	13	15
Placement Technique	Pump	Pump	Pump	Pump	Pump	Pump
Tremied/ Dropped	Tremied	Dropped	Tremied	Dropped	Tremied	Dropped
Vibration	No	No	Yes	Yes	No	No
Average Rating	0.4	1.8	1.2	2.7	0.7	1.0

1.87	1.33	1.80
1.80	1.40	1.8
1.60	1.13	1.53

Figure 12.9 Average surface rating by location on wall.

or other obstacles. In these cases, one should limit the free-fall height so as to limit the energy of the concrete as it falls. As concrete free-falls from greater heights, it can pull air voids deeper into the concrete already in place. If the free-fall height cannot be limited, then the placement rate can be slowed to again reduce the amount of entrapped voids being pulled into the concrete. However, the mixture properties must be designed and proportions established with these placement limitations in mind. Other solutions for removing the influence of free-falling concrete (in order to maximize surface aesthetics) is to pump the concrete from the bottom of the form using specially designed ports constructed into the formwork.[3,7] It should be noted that when casting walls using this method from the bottom up, the formwork must be able to withstand the pressures exerted, as the concrete at the bottom of the forms will continuously be fresh and highly fluid until placement is complete.

Additional Considerations for Casting Vertical Elements

- One should ensure that forms are watertight and level to avoid spilling and grout leakage at the seams.[2]
- SCC is not a solution for poor form quality—ensure that formwork surfaces are clean and in good condition, especially if a high-quality surface finish is desired.
- Formwork pressures. A review of the proceedings of the various conferences and symposia covering SCC will show that significant work has been conducted in the area of SCC and formwork pressures by many researchers. Current efforts are underway to create predictive models and update codes and recommendations. Some of the most recent developments include the following:
  - o   DIN[8]
  - o   RMC Research and Education Foundation[9]
  - o   Khayat el al.[10]
- Formwork pressures and the resulting requirements for form construction are a significant safety concern. Therefore, for specific guidance on formwork construction when using SCC, one should refer to the applicable codes and guidelines available.
- Drop heights. ACI 237 references a successful project with dropping heights of up to 5.8 m. The RILEM report on casting SCC references dropping heights of 8 m, with more normal heights of 1–3 m. When casting wall panels horizontally, particularly if the panels are architectural in nature, one should limit the drop height and cast as close as possible to the form in order to avoid segregation and color differences on the panel.
- Is vibration ever appropriate in SCC? Generally speaking, SCC can be vibrated if necessary. However, the purpose of the vibration in SCC must not be to ensure consolidation—it can only be meant for aesthetic purposes. This is somewhat of a gray area, particularly when using lower-slump-flow SCC mixtures. For example, the author knows of more than one prestressed double-tee producer using a slump flow in the 500–600 mm range. During the casting of these elements, the SCC mixture is allowed to flow into and fill the stems and the deck without consolidation effort. However, at the very end of each section, a poker vibrator is inserted briefly to ensure consolidation at the ends. Is this mixture considered SCC? In 99% of the element it is, but in 1% it is not. These are the areas where strict definitions and practical applications sometimes

can run counter to one another. This topic is currently being examined within the ACI 237 committee in order to provide some guidance on how to specify performance in this type of situation. There may be other situations, such as lightly sloped formed surfaces, or the bottom corners of doorways in vertically cast walls, where some light vibration or tapping is helpful, the purpose in these cases being to remove entrapped surface voids. It was shown above that simply applying vibration does not ensure a good surface finish—in some cases, the surface finish may be worsened. It is recommended that if vibration is being considered, a mock-up section be tested and the vibration techniques be evaluated.

- Can SCC be vibrated without segregating? The most appropriate answer is: it depends. Two of the things it depends upon are the mixture proportions (particularly the resulting mixture viscosity) and the intensity of vibration. To demonstrate this relationship, a simple experiment was conducted.[11] A total of six SCC mixtures were run in two groups of three, using Portland cement only. Group 1 had a paste volume of 35% and a water content of 240 kg/m^3. Group 2 had a paste volume of 30% and a water content of 170 kg/m^3. Within each group, the three mixtures were run with VMA dosages of 0, 260, and 520 ml/100 kg. For each mixture, the slump flow and the traditional slump were measured initially without vibration. Then the slump and slump flow for each mixture were measured after being subjected to vibration. The vibration was applied by filling the slump cone and submerging a pencil vibrator into the cone for 10 seconds. A third test was run applying the vibration for 20 seconds. As vibration was applied, aggregate particles settled in the cone and formed a pile in the center, and both the slump and the slump flow decreased as the segregation worsened. Figure 12.10 shows the decrease in slump by mixture as the vibration time increased. Note that as the water content was decreased and as the VMA was increased, less aggregate settled, as indicated by a smaller decrease in the slump measurement (which means fewer coarse aggregate particles piling up). This data is meant to show the interrelationship between mixture proportioning and application of vibration. Some vibration may be appropriate in certain situations, but it should always be tested first. Additionally, if one believes that vibration may be necessary in order to achieve an acceptable surface finish for example, then the concrete lift height should be minimized so that the duration of vibration required to allow an entrapped air pocket to escape is relatively shorter. Less viscous mixtures are more responsive to vibration than highly viscous mixtures.[12] In all cases, if vibration is being considered, the SCC mixture must be developed with sufficient viscosity to ensure that segregation is minimized when the vibration is applied, while also resulting in the appropriate surface finish. This is another example of two competing requirements that must be kept in balance.

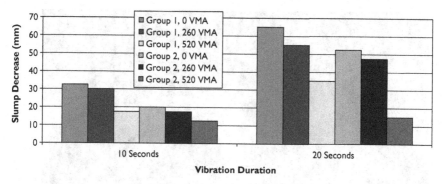

Figure 12.10 Slump Decrease due to Aggregate Settlement as Vibration Is Applied to SCC.

Casting Horizontal Elements

When casting SCC onto horizontal surfaces such as slabs on the ground, the placement crew should remember that the concrete will flow along the path of least resistance. For example, the use of an SCC mixture does not permit the placement crew to build the thickness of a slab to grade as is normally done when casting slabs. In order to cast the entire slab area at the correct depth, the concrete delivery logistics must be organized in such a way that sufficient volume is delivered to fill the area.

A large slab can be divided into smaller sections so that the area, concrete volume needed, and logistics timing can all work together as seen in Figure 12.11. In this particular project, there were several hundred meters of slab to be cast to the right and behind the photographer. In the foreground, we see a bulkhead installed to create a smaller area. Multiple smaller areas were formed, and a casting program was used that staggered the areas in such a way that previously cast sections became the side formwork for uncast sections.

If a section to be cast is large, the workability-retaining properties of the SCC mixture become more important. If the mixture begins to stiffen too quickly, then one may have concrete flowing over a previous layer, as can be seen in Figure 12.12. If this occurs, whether in a slab, wall, or other element, the two layers of concrete must be meshed together by vibration or other means.[2]

Once the SCC has been placed, it should be exposed to the least amount of agitation possible. This is not to say that the mixture (when properly proportioned) cannot survive the vibrations from a normal jobsite, but rather that concrete should not be subjected to laborers walking through it and dragging pump lines or other equipment and tools through it. Figure 12.13 shows SCC being used to cast a replacement slab by pump. One man is moving the pump line and four other laborers are walking through the concrete. The results of the project were good; however, there were too many

Figure 12.11 Large Slab Area Divided into Smaller Sections.

Figure 12.12 SCC Flowing over Previously Placed and Stiffening Concrete.

Figure 12.13 Large Crew Size Placing SCC.

laborers moving through the concrete. Additionally, one should recognize that the crew size on this project was not chosen to take full advantage of the labor-saving benefits of SCC.

In many areas, special SCC mixtures are being used to cast slabs. In these situations, the concrete producer and contractor should thoroughly discuss the benefits of the mixture to be delivered and the contractor should modify placement and finishing techniques in order to take full advantage of these benefits. For such projects, a single laborer is used to level the top surface of the slab using a specially developed tool. This tool looks like a section of small-diameter pipe with handles (Figure 12.14). It is used to tap the top surface of the slab lightly in order to create a flat, level surface. The single laborer walks through the slab in one direction, tapping the surface, and then once more in the perpendicular direction, again tapping the surface. Once the slab has been leveled, a curing compound is applied and no more further manipulation of the slab occurs. From the author's understanding, this technique is generally used for indoor slabs that will receive surface covering such as tile or carpeting.

Figure 12.15 shows the surface of a slab cast with SCC and finished with a hard-troweled surface. Similar to conventional slump concrete, concrete practitioners should recognize that some SCC mixtures, particularly those

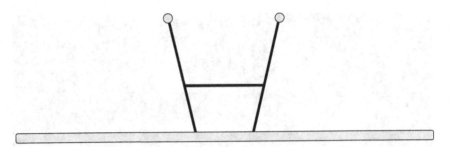

Figure 12.14 Tool for Leveling the Surface of Slabs Cast with SCC.

Figure 12.15 Hard Troweled Surface of Slab Cast with SCC.

with relatively low wv/pv ratios, will be more "sticky" and difficult to finish. An SCC mixture that will be finished in this way should be proportioned and developed with this specific application and finishing requirement in mind. In addition to the stickiness, the stiffening and setting times can be controlled through the use of chemical admixtures to provide the appropriate time windows for the placement and finishing crews.

SCC mixtures are regularly used in applications, such as prestressed elements, where a broomed or raked surface is desired. If a broom or rake is pulled across the surface of an SCC mixture too early, the profile of the ridges will likely be insufficient, as the concrete will simply flow back to a level condition. The timing of the finishing process relative to the stiffening rate of the concrete mixture is critical in this case. It is the stiffening or increase in yield stress (decrease in slump flow), and not necessarily the setting time, that will permit the concrete to hold a broomed or raked surface. This application again requires that the SCC mixture and the process timing be considered together so that the optimal balance can be achieved for the project. Similar to conventional concrete, the correct choice of mixture proportions and chemical admixtures can be used to create a wide range of stiffening and setting profiles for SCC mixtures.

Curing of SCC

In many cases, the bleeding characteristics of SCC are relatively low compared with those of conventional concrete. However, the bleeding characteristics are a function of materials and mixture proportions, and a wide range of bleeding levels is possible. This must be considered during the proportioning and testing phase in order to prevent surface crusting and plastic shrinkage cracking. Where this possibility exists, the use of fog-misting equipment or other standard practices used for conventional concrete can be employed. ACI 237 recommends that the established guidelines for curing conventional concrete, such as those found in ACI 308R, should be followed with SCC.[13]

Summary

In general, SCC can be transported and placed with the same methods used for conventional concrete. However, some considerations with SCC should be taken into account, such as batch size and mixture properties versus the terrain over which transport vehicles must travel, the relationship between placement technique, energy, flow direction, and mixture properties, and the relationship between the mixture properties and the finishing and curing requirements. Additionally, the contractor should consider how to adjust the placement process to take full advantage of the benefits that SCC presents, but also to ensure that the planned techniques will not inhibit the SCC from performing satisfactorily.

Chapter 13

Troubleshooting SCC Performance

Introduction

A new technology, technique, or piece of equipment is incorporated into an existing process and, inevitably, an unexpected result occurs. When this happens, one can respond in one of three basic ways:

1. Immediately stop using the new technology and go back to the old process with no intention of trying the new technology again.
2. Continue using the technology in the same way as before and hope the unexpected result (if it was negative) doesn't happen again.
3. Investigate why the result occurred so that it does not happen again (if it is negative) or learn how to repeat it (if it was positive) with the intent of fully implementing the technology.

Of these options, the first is not really an option when the technology has real value to the user. Likewise, blindly ignoring a problem in the hope that it will go away is not an option either. In the case where the technology has value, the best response is to move forward and refine the technology so that the desired results are consistently achieved.

Anyone implementing an SCC program will invariably experience an unexpected result at some point. Being prepared to investigate an unexpected result is essential. This section will outline the general process of troubleshooting, discuss several troubleshooting case studies, and finally present a practical guideline for troubleshooting common SCC problems.

The Troubleshooting Process

A review of the troubleshooting process may seem elementary to some. However, the more one is involved in field technical service issues, the more one realizes how often conclusions are drawn and decisions made based on very thin evidence, ultimately resulting in a still unresolved question. When an unexpected result occurs, a systematic approach following the basic scientific method will most efficiently determine the cause. The following are the general steps of the troubleshooting process:

1. Collect all quantitative and qualitative data related to the result. This includes but is not limited to concrete production and testing data, placement data such as pump pressure changes during placement, and observed mixture performance differences reported by laborers at the casting site. Do not make judgments on the importance of individual pieces of data at this point. It may be that information that is seemingly unimportant at the beginning of a troubleshooting process may be very helpful in drawing conclusions at the end of the process.
2. Ask the question "What are all of the potential causes of this problem?" and create as thorough a list as possible of all potential causes. This is important because, as much as one would always like there to be a single cause, in some cases there may be multiple factors leading to the unexpected result.
3. From the data collected and the potential causes, form one or more hypothesis to explain the occurrence.
4. In the case where insufficient data have been collected during production, conduct an experiment to validate the hypothesis; the experiment can be as simple as making adjustments to the field processes used and observing the results, or it can be as detailed as running a controlled laboratory experiment. In cases where sufficient data have been collected, an experiment may not be needed and a reasonable explanation and recommendation can be made. In either case, a systematic approach should be followed, unless of course the proverbial "smoking gun" is immediately found—which is quite rare when troubleshooting concrete performance issues.
5. Confirm or reject the hypothesis and if necessary create a new one based on the latest data.

Collecting Data and Information

Whether an experiment is run or not, data are required to solve any problem. The quality control process discussed earlier prepares one for troubleshooting unexpected results. The quality control data collected, however, need to be diverse enough to allow characterization of the problem. This being the case, all information is potentially useful, so the investigator should collect all that is available, whether it is quantitative or qualitative. This is analogous to the process of putting together a picture puzzle. If one has only a single piece of the puzzle, it is difficult to describe the picture. The more pieces collected, the greater the likelihood of fitting them together and clearly revealing a section of the whole picture.

At the beginning of an investigation, when data are not immediately available, many questions of a general nature are asked, such as "Has this problem ever happened before?" If the answer is "yes," then one should drill down into that history of when, how, and why. One also needs to know how consistently a problem occurs. Does it happen on every batch or during every pour? Does it happen at a particular time of the day or year, after it rains, or when it is dry? Does it occur only intermittently? Intermittent, inconsistent problems are often the most difficult to solve. When a problem occurs consistently, that means its cause is more likely to be consistent and ultimately somewhat easier to find. The questioning process helps to frame and create problem boundaries. Once the boundaries have been created, more specific questions, such as "Has there been a new delivery of raw materials (cement, other powders, fine aggregate, coarse aggregate, admixture) recently?" can be asked. Similar to an automobile mechanic or a physician, the investigator uses questioning in a process of elimination. If potential causes can be eliminated via questioning with a reasonable degree of confidence, the process of finding the ultimate cause becomes smaller and more manageable.

Like conventional concrete, troubleshooting may be required for SCC in either the fresh or hardened state, so it is helpful to have some knowledge of the variables that influence certain SCC properties. It is also useful to be experienced in concrete production equipment and processes, as well as placement and use. Having diverse experience to draw from provides the investigator with multiple perspectives from which to look at a problem.

Troubleshooting Fresh SCC Properties

The fresh properties of SCC are critical to its successful use. When an issue arises, one of the first steps is to collect the batch ticket and confirm that the mixture was batched correctly. This should include confirming that the correct aggregate moisture adjustments were made. This section will review some of the more common SCC fresh performance problems. These troubleshooting topics all assume that an SCC mixture has been developed with good properties and is currently in production.

The Slump Flow Does Not Increase When More HRWR Is Added

In some instances, after a mixture has been proportioned, producers have reported an inability to increase the slump flow regardless of how much HRWR is added. Two primary things must be remembered in this situation: first, the HRWR increases the fluidity of the paste fraction only, and, second, the relationship between the paste volume and aggregate characteristics determines the concrete mixture fluidity. If the paste volume of the mixture is too low, it may be susceptible to swings in aggregate gradation that cause excessive interparticle contact and inhibit the concrete fluidity. In this scenario, addition of more HRWR only creates an overly dispersed paste, resulting in mixture instability. Additionally, if the mixture is originally designed with a very low water content and a high dosage of HRWR, one can reach the point where the mixture cannot be further dispersed with additional admixture. This mixture may therefore be influenced by erroneous moisture calculations. In either of these cases, it is a mixture proportioning issue. These mixtures were developed with insufficient robustness. Robustness should be evaluated during the mixture development process so that the SCC is able to provide the desired performance when normal material variations occur.

The Slump Flow Is Increasing from the Plant to the Jobsite

The dosage of HRWR required to achieve a certain slump flow is determined during the mixture development process. If, during production, the dosage increases substantially above the design, this could result in an increasing slump flow during transit. The causes of a dosage increase include material changes, temperature changes, and inaccurate moistures. The most common is erroneous moisture values. If the free moisture is overestimated, a lower water content will result, which may require longer mixing for the mixture properties to stabilize (see Chapter 10). If this extra time is not accounted for, then the slump flow measured at the plant may be lower than that measured at the casting site.

A Previously Stable Mixture Is Now Showing Signs of Segregation

Segregation is the most frequently reported fresh SCC problem. If a mixture is segregating, one should notice the type of segregation. Some mixtures will show segregation immediately, while others will show delayed segregation. If the mixture segregates immediately (<30 seconds), this is representative of a lower-viscosity paste, which is typically caused by excess water. At this point, the water content and aggregate moistures should be checked. A delayed segregation can be seen when no immediate segregation occurs and the mixture is left to sit undisturbed in the wheelbarrow. Within 1–2 minutes, the top of the concrete will be covered with a discolored paste or bleed. This delayed segregation is often associated with a higher-viscosity paste and an increase or overdose of HRWR. If it is accompanied by a slump flow that exceeds the maximum, then one should reduce the HRWR dosage and evaluate. If it is accompanied by a slump flow that is within the normal range, then one should check that moistures are correct; if they are, then a material change may have occurred and a mixture proportioning adjustment may be necessary.

Troubleshooting Hardened SCC Properties

An improved formed surface finish is one of the most highly anticipated benefits of using SCC. From the author's experience, the main reported problems with SCC in the hardened state are surface finish imperfections. This general class of problem can be made manifest in several ways, each of which will be discussed in the following sections.

Sand Streaking and Bleed Channels

These occur in cast elements, where a bleed water trail can be seen running vertically at the formed surface. The trail is usually empty of paste, and only aggregate (primarily fine aggregate) can be seen. This behavior requires excess water to be present, either within the mixture, which can then move to the form surface, or inside the form prior to concrete placement. If the sand streaking is caused by excess water in the mixture, the SCC will appear to be close to segregation and it will likely have a very low (fast) T50 time. In one case in which the author was involved, a mock-up wall was cast by pumping. Before filling the pump with SCC, a highly fluid mortar was run through the pump as a line slick. Some of this mortar was still left in the pump line at the start of pumping and discharged into the form. As the concrete was placed in the center of the form, the very fluid mortar was forced ahead of the concrete to either end of the mock-up wall, resulting in bleed channels on each end. Sand streaks can also result from long delays in concrete setting where the mixture stays fluid for too long a period of time after placement.[1]

Surface Voids

Voids on the as-cast surface of a vertical element can be caused by a number of different factors, including the mixture properties, the form release agent, and the placement technique. In some cases, an analysis of the characteristics of the voids is helpful. Figure 13.1 shows an example of relatively large irregularly shaped voids and Figure 13.2 shows an example of relatively smaller, round, symmetrical voids.

Irregularly shaped voids can be entrapped at the surface owing to poor flowing properties of the SCC mixture or poor placement techniques. The poor flowing properties can be a result of too low a slump flow, too high a viscosity, or insufficient paste. Rapid placement and free-fall of SCC can also draw air into the mixture, and this air is then trapped at the form surface. A review of the placement techniques, mixture performance targets, and the proportions may be necessary, depending upon the frequency of this occurrence. Good-performing SCC will generally not result in these types of surface voids, even when super-high-quality surface finishes are not targeted.

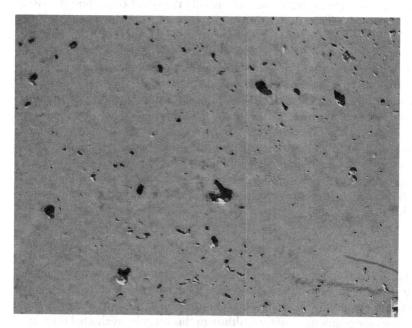

Figure 13.1 Irregularly Shaped Voids.

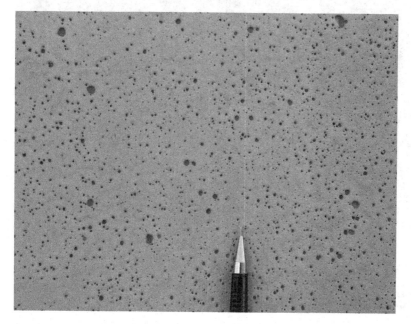

Figure 13.2 Round Symmetrical Voids.

Small round symmetrical voids are usually the result of air bubbles rising to the concrete surface. In the author's experience, the primary cause of this is instability. Figure 13.3 shows the top surface of a sample of fresh, unstable, air-entrained SCC. Notice the large number of symmetrical bubbles at the surface. If this concrete is placed into a form, the bubbles can be pushed to the form's surface and leave small round voids.

In some instances, the surface voids are not apparent upon de-molding, but are revealed when the elements are sand-blasted or acid-washed. This may be more a function of the form release agent rather than any difference in the SCC mixture. It has been reported that water-based form release agents can in some cases encourage the formation of a thin paste layer at the form–concrete interface.[1] This thin layer hides the voids, and when it is removed the voids appear. If these voids appear, the mixture should be evaluated for stability.

Pour Lines

Similar to conventional concrete, lines showing on the face of an element can occur if two successive placements do not mesh together (Figure 13.4). In the worst of these situations, cold joints can occur; most often, however, these pour lines are cosmetic in nature, although this must be evaluated on a case-

Figure 13.3 Unstable Air-Entrained SCC Mixture.

Figure 13.4 Pour Lines.

by-case basis. Pour lines are usually caused by a combination of placement delays and either a thixotropic characteristic or inadequate slump flow retention of the SCC mixture. Mixtures with low water/powder (w/p) ratios and high powder contents may have more of a tendency for thixotropic behavior. For this reason, SCC must be tested for time-dependent flowability properties during the mixture development phase. If pour lines occur, the mixture should be retested for thixotropy and for slump flow retention. If these properties are now present but were absent during the development phase, then one can investigate what has changed. Higher-than-expected ambient and concrete temperatures can influence this behavior and should be recorded. The temperature of the forms themselves, particularly if placement occurs outside, can also contribute to rapid stiffening of the placed concrete.

Honeycombing or Large Voids

These usually occur owing to poor passing ability and/or a lack of mixture fluidity. The mixture should be reevaluated for both of these properties as well as for slump flow retention. If the coarse aggregate grading has changed such that a higher amount is now being retained on the larger sieves, a mixture proportioning adjustment may be required to lower the coarse aggregate content (via the CBI proportioning methodology).

Gaining Insight from Both Fresh and Hardened Properties

It may be possible to isolate the cause of an unexpected result by evaluating certain fresh and hardened properties together. In contrast to conventional concrete, with SCC, more information on the rheological characteristics of the mixture is collected. In particular, certain test methods provide a reasonable estimate of the mixture's viscosity. We know that the w/p ratio of a mixture is a primary variable affecting mixture viscosity, while the water/cement (w/c) ratio is the primary determinant of compressive strength. Knowing this, we can use a viscosity and a compressive strength measurement to isolate water fluctuation in a mixture.

Figure 13.5 was developed based on a set of published data.[2] It shows the average T50 time versus the average 24-hour compressive strength of two high-strength SCC mixtures being evaluated for prestressed/precast applications. Each of the data points is the average compressive strength for a single batch of the mixture represented. Eight batches of SCC Mixture 1 and five of SCC Mixture 2 are plotted. Notice the correlation between 24-hour compressive strength and the T50 time. As the T50 time increases, the compressive strength also increases. If found during troubleshooting, this type of relationship could indicate that unaccounted-for moisture is the cause of the mixture variability.

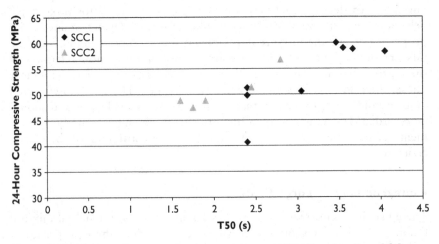

Figure 13.5 24-Hour Compressive Strength versus T50 Time for Two SCC Mixtures.

SCC Troubleshooting Case Studies

This section presents five case studies where SCC was used and unexpected results occurred that required investigation and recommendations for improvement.

Case Study I

SCC was specified for use in cast-in-place architectural exposed walls for an art museum expansion. A large well-respected ready-mixed concrete producer was supplying the SCC for this project. This producer had previously supplied SCC for other projects and therefore had some performance history with SCC mixtures. On this project, the initial quality control testing requirements were only for slump flow.

On one particular pour, the producer's SCC expert was unavailable for the start of the pour. By the time he arrived on site, three truckloads of SCC had been delivered and placed (by pump). As the next batch arrived and was tested for slump flow, the representative noticed problems with the SCC mixture. The mixture looked "wet" and was flowing faster and more like water. The representative contacted the concrete batch plant, reduced the water content, and increased the HRWR dosage. With this change, the mixture performance returned to what he considered normal. Three full truckloads, however, had already been placed into the wall. At the end of the pour, when the wall was filled, the contractor had to remove several inches of watery paste from the top of the wall because the first batches of SCC had segregated. In subsequent conversations with the contractor, he indicated that the concrete from the first three truckloads flowed around the walls much faster than in previous placements. Through review of batch records and interviewing plant personnel, the concrete supplier determined that the aggregate moisture contents were not compensated for appropriately during batching. By not making the necessary adjustments, approximately 20 liters of water per cubic meter of concrete were unintentionally added to the mixture. The confusing part of this scenario, however, was that the slump flow level tested on site did not dramatically change.

The contractor and supplier decided that no further placements would occur until everyone was in agreement with how the concrete mixture was to perform. This meant that the contractor foremen, the testing laboratory personnel, and the technical representatives from the concrete and admixture suppliers were to be in agreement as to what constituted acceptable SCC.

To accomplish this, a meeting and training exercise was held at the concrete production plant. The exercise consisted of batching the originally designed mixture, testing it for slump flow, T50, VSI, and J-ring, adding water in 10 l/m^3 increments, testing again and repeating up to a cumulative total of 30 l/m^3. This exercise was designed to demonstrate to the site personnel the influence of extra water, what good and bad SCC look like, and finally to establish a more complete set of quality control parameters. Table 13.1 shows the results of the exercise, with the resulting change in each parameter as water was added to the batch. Note the predictable changes to slump flow and T50 times: as water is added, the slump flow increases and the T50 time decreases. The VSI and J-ring, however, showed no change until the final water addition. During this exercise, it was noted that the sample in the wheelbarrow, taken from the final treatment with 30 liters of additional water, was clearly segregating, with aggregates settling to the bottom and a watery paste rising to the top. Input from the testing personnel indicated that this sample resembled the initial three batches that were delivered to the site and placed, especially when looking at the sample in the wheelbarrow. This occurred even though the VSI rating of the slump flow paddy was 1. It is interesting to note that the original VSI test relied not only on a visual examination of the slump flow paddy alone, but combined it with a visual examination of the concrete in either the mixer or in the wheelbarrow for exactly this reason.[3] Sometimes, segregation cannot be seen in the slump flow patty because of a lack of depth; however, the same mixture observed in a sample with depth (in a bucket, wheelbarrow, etc.) now has the opportunity for solid particle settlement.

The final unanswered question was with respect to the slump flow values on site being acceptable. Once all of the initial testing had been completed, a second slump flow test was performed on the final sample in order to demonstrate the impact of sampling procedures on the test values of segregated SCC mixtures. For all of the previous slump flow tests in this exercise, a sample was taken from the wheelbarrow with minimal remixing. This final time, however, the technician purposefully tried to remix the concrete and obtain a representative sample by taking material from the bottom of the wheelbarrow. The result of this test was a slump flow of 650 mm, versus the 735 mm slump flow when a sample was taken from the sample without significant remixing. This highlights the potential variability of fresh property tests when a mixture is highly segregated. Obtaining a representative sample of mixtures in this condition is difficult—hence the need for conducting other tests in addition to the slump flow.

Table 13.1 SCC Parameters as Water Is Added to a Single Batch*

	As-Batched	10 l/m³	20 l/m³	30 l/m³
Slump Flow (mm)	570	620	660	735
T50 (s)	2.3	1.9	0.9	0.7
J-Ring Difference (mm)	25	15	25	75
VSI	0	0	0	1

*Water values stated are cumulative.

Final Resolution/Recommendation

The placement crew's observation of faster-flowing SCC, the testing personnel's confirmation that the mixtures looked segregated, and the batch testing results all tended to confirm that the producer needed only to more accurately control moisture contents. This was the action taken by the producer, and with this adjustment the rest of the placement occurred successfully.

Practical Lesson Learned

- Before the start of an SCC project, a pre-job discussion between the concrete producer, contractor, and testing agency where all parties agree on the appropriate SCC parameters and testing responsibility should be conducted.
- Always ensure control of aggregate moistures with SCC.
- Those without SCC experience should receive training, including hands-on witnessing of good and bad SCC.
- Do not ignore concrete that looks as if there is a problem, even if the "test results" are within specification.
- Segregated mixtures will have greater variability in test values than stable mixtures, and therefore a single test, such as the slump flow, by itself is not adequate to ensure good SCC. For quality control, always perform the VSI and T50 tests along with the slump flow. These tests do not guarantee good performance; however, they do increase the likelihood that a problem batch of concrete will be caught before it is placed.
- The sampling procedure outlined in ASTM C 172[4] states that production samples of concrete obtained for testing should be combined and remixed with a shovel to the minimum degree necessary to ensure uniformity. Technicians are trained to remix samples prior to filling test molds or equipment. When an SCC mixture is unstable, this could result in misleading results. Technicians should be instructed to take testing samples that are representative of the concrete in the receptacle and not try to "make it uniform" when it is not.
- A producer should always run a robustness test on the SCC mixture being used. This can be accomplished similarly to the testing presented in Table 13.1: batch the mixture, test it, and add water to it, taking various measurements after each addition. Record any qualitative change that is noticed, in addition to the test results. This record can serve as a living quality control document/manual as the project moves forward.

Case Study 2

A precast producer was using SCC, delivered and placed by mixer trucks, to cast integrally colored wall panels. The panels would serve as structural components as well as providing the architectural finish for the building. The panels were cast on horizontally positioned steel forms located outside in direct sunlight. Multiple panels were positioned in a direct line and the casting process consisted of filling each panel in order, moving from one panel to the next. After de-molding, the formed face of the panels received light sand blasting to achieve the final desired finish.

It was after the sandblasting process that the producer was noticing problems. The sandblasting revealed random areas containing small clusters of voids ranging in size from 2 to 10 mm. The surface appearance was considered unacceptable owing to the presence of these areas. In an attempt to solve the problem, the producer spent considerable time making multiple adjustments to the SCC mixture and to the admixture dosages, none of which were successful. The author spent a day witnessing the preparation, batching, and casting process. No issues were evident in the preparation and batching process; however, several items were noticed during the casting process.

The steel forms were prepared in the morning, and casting began close to noon, during the hottest part of the day. Prior to concrete being placed into the forms, a surface temperature on the forms was taken and found to be nearly 49 °C (120 °F). Each mixer truck held a volume of concrete that could fill more than one panel. Once a panel was cast, the remaining concrete would be held in the truck until the placement activities on the previous panel were completed; then casting would begin on the next panel. However, concrete was left not only in the truck but also in the chute. On this occasion, a panel was completely filled, and the chute was then raised level and swung over the empty panel to be cast next. This was done to avoid concrete being spilled on the ground around the casting area. The SCC in the chute continued to run out and drop from a height greater than 1 m above the form. As concrete fell onto the form, some would splatter and shoot concrete to other areas around the 49 °C steel form. The concrete sat on the form for 5–10 minutes while the previous wall panel was completed. Over this time, enough concrete had flowed from the chute that a 1 m diameter area had been covered with SCC. The concrete sat undisturbed on the steel form, and began to stiffen and harden in place. When casting finally started on this panel, the fresh concrete flowed over the top of the hardened material, and no attempt was made to mesh the material together. The location of this larger area was marked on the form and the following day the panel was sandblasted. In this area where the concrete had been dropped and allowed to sit on the hot form, a curved line of small voids was revealed.

Final Resolution/Recommendation

The casting of architectural wall panels requires great attention to detail. Seemingly small items can reveal themselves once sandblasting or other final treatments have been done. It was recommended that the producer ensure that each panel be cast completely without interruption, particularly during the hot summer months when the forms reached high temperatures. It was also recommended that the concrete never be allowed to drop into the forms from a height greater than about 300 mm.

Practical Lessons Learned

- Always be aware of the forms' surface temperature when casting elements outside.
- The practice of positioning the chute over the next form and not allowing concrete to spill over the casting area is a good and acceptable practice. When done routinely and without forethought, however, good and acceptable practices can still sometimes be the cause of undesired results.
- When an unexpected or undesired result occurs, fully consider all of the possible causes. Rather than evaluating the entire process, the producer immediately started making changes to the mixture that wasted time, effort, and resources.

Case Study 3

A large, well-respected precast concrete producer was using SCC to cast prestressed bridge girders. The girders were nearly 2 m deep, with dense prestressing strands running through the bottom. Their shape was the typical I-beam shape. As the bottom portion containing the strand narrowed into the tall web section, it ran on an inward slope from bottom to top. SCC placement was by crane and bucket, and multiple placements were required to fill each girder form. Concrete was delivered to the bucket by mixer truck from a concrete plant located on the property.

The problem experienced by the producer was a concentration of very weak paste along the face of the inward angled section. Evaluation of this area showed that the surface appeared to contain extremely high levels of air, almost to the point of resembling foamed paste. These concentrations of foam did not appear anywhere else on the beams and were appearing at a reasonably consistent rate. The State Department of Transportation was concerned that this issue was indicative of some other problem in the mixture or in the girder that was not visible. Cores were taken from strategic locations and no compressive strength problems were found. Even though air contents were normal, the issue was assumed to be a problem with the concrete mixture, since the high concentration of air was present within the concrete. However, none of the mixture or admixture adjustments made resulted in the elimination of these weak surface paste areas.

When the author was contacted regarding this situation, one of the first questions asked was whether the mixture was stable. When some SCC mixtures are highly unstable, foam will rise to the surface. If an unstable SCC mixture was placed into these bridge girder forms, the unstable foam could rise and be trapped against this inward angled surface. According to the quality control manager, however, there were no indications of segregation during quality control testing or during placement. Therefore, the entire process of batching, delivering, testing, and placing the concrete was witnessed. No issues were evident until the concrete was about to be placed. Before loading the bucket with concrete from the mixer truck, a member of the placement crew sprayed a large quantity of form release all over the inside and outside of the bucket. The composition of the form release was unknown. This was a typical process used to reduce build-up on the bucket. The amount sprayed, however, was excessive and the excess could be seen in concentrated puddles along the bottom edges of the bucket. When the SCC mixture, which showed no signs of instability, was poured from the truck into the bucket, the concrete and the excess form release swirled together. As the bucket was filled, a thick band of stable foam was formed along the entire edge of the bucket. The size of the band was estimated to be 75 mm by 75 mm. This foam was not composed of the typical unstable large bubbles that occur when an SCC mixture is segregating. Rather, the bubbles were quite stable and did not disappear as more concrete was placed into the bucket. The bucket was then lifted into position and the concrete placed into form. When the bucket was returned to be refilled, it was completely empty, meaning that both the concrete and the foam, previously in the bucket, were now in the form. It was presumed that the foam generated as the bucket was filled would likely rise in the form with the level of the concrete and be trapped against the inward angled steel form surface.

Final Resolution/Recommendation

It was recommended that the placement crew reduce the amount of form release sprayed into the empty bucket so as to minimize the generation of foam along the bucket side. If this could not be done, then the foam that did develop along the side of the bucket should be removed and not deposited into the form with the concrete.

Practical Lessons Learned

- Activities that are unrelated to batching, mixing, and producing fresh concrete can present themselves as concrete mixture problems. In this case, the presence of foam was assumed to be the result of a concrete mixture problem, but was not.
- Excess form release may generate and stabilize foam as it is swirled with concrete during placement.
- Be aware of the interaction/performance of the form release agent being used. It is quite feasible that this same type of phenomenon could occur when excess form release is present in the form itself.

Case Study 4

A highly fluid SCC mixture was specified for use in filling a large open horizontal space underground. Concrete was transported to the underground casting site via a concrete bucket and pumped through tubes placed horizontally to the point furthest into the horizontal excavation. As the concrete was placed, the tubes were retracted and the horizontal space was slowly filled with SCC. Because of the slow placement process and because it was a mass placement, a combination of Type I Portland cement and slag cement was used in the SCC mixture. The slag cement content accounted for 50% of the total cementitious material. The project was successful and resulted in time and labor savings, as well as providing the contractor with a significantly safer method for placement.

Based on the success of this first project, the same engineer specified SCC for use on a second project located more than 1000 miles (1600 km) away. The identical mixture proportions from the first project were prescribed for use on this second project. In this case, however, the mixture was proportioned with 50% Class C fly ash because this was the only locally available pozzolan. This SCC mixture performed very differently from the original mixture. It stiffened so rapidly during placement that the concrete pump was plugged, which caused delays and serious problems for all involved.

Final Resolution/Recommendation

A laboratory testing program examining the interaction of the powders and admixtures was conducted. Based on this experiment, it was found that the amount of Class C fly ash had to be reduced in order to eliminate this rapid stiffening behavior. At this point, adjustments to the mixture proportions were allowed, a revised mixture was developed, and the project continued successfully.

Practical Lessons Learned

- Combinations of different powders will behave very differently with respect to time-dependent flowing properties.
- It is nearly impossible to take a full set of SCC mixture proportions from one geographic location to another and achieve identical results.
- SCC should be specified by performance requirements and not by mixture proportions.

Case Study 5

An architectural precast producer was using SCC to cast wall panels with an exposed aggregate finish. Concrete was mixed and delivered to the casting bed via a concrete bucket. To achieve the required finish, a high volume of coarse aggregate was required. The coarse aggregate being used had a relatively narrow particle size distribution, with over 40% being retained on the 4.75 mm sieve. The producer was experiencing three primary problems: it was not possible to achieve the level of fluidity desired to optimize placement, the mixture was becoming unstable, and severe workability loss was occurring to the extent that the concrete had to be vibrated out of the bucket. The problems experienced and their potential causes were discussed and a laboratory investigation was conducted to isolate the causes and recommend solutions. Table 13.2 shows the problems and potential causes that were evaluated.

When evaluating the slump flow of the original mixture used in production, a high concentration of coarse aggregate particles, particularly in the center of the slump flow paddy, was immediately apparent. In an SCC troubleshooting scenario, if this type of slump flow is seen, one can likely assume that the paste volume is not sufficient to promote proper flow. If excess HRWR admixture is added to a mixture like this, one will likely see minimal increase in slump flow, but may see the paste begin to segregate from the aggregate. This paste separation will typically occur at the outside edge of the slump flow paddy.

A laboratory experiment was conducted to evaluate mixture proportion adjustments and raw material substitutions. When the paste volume of the mixture was increased (which also required a decrease in aggregate volume), the slump flow dramatically improved, as did the mixture stability. In the case of material substitutions, the cement being used was found to be the primary cause of the rapid stiffening, although the paste volume increase also resulted in a slight improvement in slump flow retention.

Final Resolution/Recommendation

The proportioning changes were not possible at this time because the exposed aggregate finish would be impacted if the recommendations were imple-

Table 13.2 Case Study 5 Problems and Potential Causes

Problem	Potential Cause
Inability to achieve desired fluidity	• The high coarse aggregate volume required for the exposed aggregate surface. • False or flash-setting cement causing the concrete to rapidly stiffen.
Mixture instability	• The high coarse aggregate volume, resulting in an inadequate volume of paste that was then overdosed with HRWR in order to achieve the fluidity desired.
Rapid stiffening	• A false or flash-setting cement.

mented. Therefore, the producer continued to use a mixture that was close to SCC consistency and applied light vibration for consolidation. To address the rapid stiffening problem, further investigation with the powder supplier was recommended.

Practical Lessons Learned

- One variable, such as paste volume, can result in several problems, including insufficient flow, instability, and slump flow loss.
- Determining the cause of a problem and then implementing the recommended solution may not always be possible. In this case, the exposed aggregate finish was more important than the ability to place the mixture more easily.
- An SCC mixture should be thoroughly developed and evaluated before its implementation.

SCC Troubleshooting Guidelines

No single table can easily outline the causes for all unexpected results when using SCC. In many real-world situations, it is highly probable that multiple issues are causing a problem or that multiple problems are occurring simultaneously, and therefore adjusting one variable may not completely resolve the issue. Table 13.3 is meant to highlight some of the typical causes, and should be used as a starting point for troubleshooting investigations.

Table 13.3 SCC Troubleshooting Guidelines

Problem	Potential Cause	Suggested Remedies or Variables to Investigate
Irregularly shaped bugholes or voids on the formed surface	• Slump flow may be too low. • Placement technique may be entrapping large voids. • Insufficient paste content. • Mixture viscosity may be too high, not allowing air entrapped during placement to escape.	• Increase the slump flow. • Limit dropping height of concrete. • Pour more slowly to match speed of concrete flow. • Decrease the mixture viscosity. • Adjust the aggregate proportions to enhance flow. • Tap or vibrate lightly during casting.
Round symmetrical voids on the formed surface	• Unstable mixture. • Excessive form oil.	• Ensure mixture stability upon completion of mixing and over time. • Reduce form oil thickness.

Continued

Table 13.3 Continued

Problem	Potential Cause	Suggested remedies or Variables to Investigate
Honeycombing	• Slump flow too low. • Inadequate passing ability due to improper aggregate and paste volumes. • Inadequate passing ability due to poor mixture stability.	• Adjust aggregate proportions. • Increase paste content. • Increase slump flow. • Adjust mixture viscosity.
Incomplete filling of the form	• If this occurs underneath a box-out, then placement from multiple directions may be the cause. • Insufficient slump flow.	• Increase flow. • Adjust aggregate proportions. • Decrease viscosity. • Slow placement rate.
Sand streaking	• The mixture is unstable. • Free water or fluid in forms prior to placement.	• Decrease the water/powder ratio. • Decrease the slump flow. • Increase the VMA dosage. • Ensure there is no excess water or fluid in forms. • If concrete is pumped, ensure that line slick is not deposited directly into form.
Lift lines	• Extended delay between successive placements. • Inadequate slump flow retention (due to material and/or temperature changes). • Mixture has thixotropic characteristic.	• Evaluate for thixotropy. • Increase water/powder ratio if possible. • Correct placement timing delays. • Increase dosage of workability-retaining admixture. • Increase slump flow.
Aggregate blocking	• Poor mix design.	• Adjust aggregate proportions. • Increase paste volume.
Excessive bleeding	• Slump flow too high. • Water/powder ratio too high. • Insufficient VMA.	• Decrease slump flow. • Decrease water/powder ratio. • Increase VMA dosage or add entrained air.
Slump loss	• Ambient or concrete temperature increased. • Inaccurate moistures, leading to a low water/powder ratio and lower slump flow. • Reactivity change of the powders in use.	• Ensure correct moistures. • Increase workability-retaining admixture dosage. • Use extended slump HRWR. • Evaluate admixture powder interaction.

Summary

Inevitably, unexpected results will occur, and this should not be a surprise. A producer or user can prepare for this eventuality by diligently adhering to a quality control program. This will ensure that data that can be used in the troubleshooting process are regularly collected. The cause of an unexpected result may not be immediately apparent, and some causes may be hidden in peripherally related activities. Therefore, when an issue occurs, a systematic approach to solving the problem is the most efficient way to handle it. Although it may be tempting to jump into performing a single test or making multiple process or mixture changes to fix the issue, many times this can lead to wasted time and effort. Troubleshooting can eventually lead to a greater understanding of one's materials and processes, as well as to new developments to ensure that the same mistakes do not happen again.

Chapter 14

Implementation and Training

Introduction

The requirement to implement an SCC program can come proactively based on internal motivation or from outside forces, usually in the form of a specification. How this requirement occurs can influence the perceived benefits and final success of the SCC program. Two examples highlight this concept. In the first case, a contractor began a project using conventionally vibrated, high-slump concrete to cast reinforced concrete walls. Upon demolding, the walls were covered with imperfections, which was unacceptable since the walls would be painted and exposed. Patching was undesirable from both an aesthetic and an economic perspective. The concrete producer suggested the idea of trying SCC. A mock-up wall was approved and the placement was successful. SCC was used for the remainder of the project and resulted in an almost flawless surface finish that required no patching. This contractor has gone on to study the use of SCC and has implemented the technology into multiple other projects. In the second case, SCC was specified by the engineer for use in the construction of a high-rise building. In this case, the contractor said he had difficulty seeing the benefits of SCC and "could have done it just as well the old way." On subsequent questioning, however, this second contractor confessed that he had not spent significant time finding ways to change his processes and take advantage of what SCC could offer him. This story is not an indictment of the second contractor but rather an example of how the motivation to use SCC can have a very real impact on the extraction of its benefits. The first contractor had a problem that was solved, he experienced the value of SCC, and went on to further study and benefit from it, while the second contractor was made to use SCC because it was specified. This second contractor was not invested in its use. The successful implementation of SCC (or any new technology) requires commitment. Without this internal motivation, the effort required to search out and take advantage of the benefits may be lacking.

Implementation

The implementation of new concepts requires change, and change requires effort. The rate, and potentially the success, of SCC implementation, will be affected by an organization's or individual's resistance to change. Resistance to change, however, is not always a result of short-sightedness or disagreement over the validity of the proposed change. Resistance can also come from being forced to change without being given the right tools or training to make it happen. Implementing SCC requires the creation and communication of both a clear goal and the path to get there. The plan itself will differ between producers, contractors, engineers, etc., but some aspects of the implementation process will be similar regardless of the group. A well conceived, communicated, and executed plan ensures that each step has been reviewed and the necessary contingency plans put in place. This chapter will review the general concepts of such a plan, knowing that detailed implementation plans are site and project specific.

Clearly Identify the Reasons for Using SCC

In this first step of implementation, a goal is set for using SCC. Organizationally, what benefits do you intend or desire to receive by using SCC? Are they solely economic in nature, quality-based, or even safety-focused? As a contractor, do you want to produce an as-cast, architectural surface that requires no remediation? Do you desire to re-deploy labor and reduce overall placement costs or construction time? As a ready-mixed concrete producer, do you want to differentiate yourself from the competition and provide a specific solution to your contractor customers? It may be that there are multiple reasons for using SCC, and some may be more important than others. Whatever the reasons, they should be as quantifiable and as clear as possible, so that they can be clearly communicated and measured. At the end of the evaluation period, you will want to know how successful you were.

Create an Implementation Team

To accomplish any goal, action is required. One does not simply say "Let's implement SCC" and expect it to happen. Someone has to be responsible for the success or failure of the implementation. Therefore, the first concrete action is to create a team that has the responsibility for thoroughly researching and understanding SCC. This team is responsible for developing the detailed implementation plan, including tool development and training needs. Preferably, the team has representation from all of the groups that will be interacting with the concrete first hand. As we have seen in previous chapters, a successful SCC program is not just a quality control responsibility—it also requires effort from production, operations, sales, etc. The team should also have a champion or leader with authority to direct the activities of the team and report its progress to management.

Assess the Organization's Current Ability to Implement or Use SCC

For a concrete producer, the starting point for this assessment is the current quality control process. Has the organization historically produced or used other high-performance concrete mixtures? If so, what quality control processes are in place, and how successful are they? One should assess the organization's ability to consistently produce its current concrete mixtures and focus attention on any limitations that may influence the production of SCC. In particular, one needs to assess the ability to control those items reviewed in Chapters 9–11. In some cases, producers have tried to use SCC at some point and then abandoned the idea for one reason or another. Does the organization have a history with SCC? Is the history positive or negative? One should learn from both the positive and negative aspects of what has been done in the past, and also from others in the industry.

Assess whether or not you are organized to take advantage of the benefits of SCC. How should labor deployment or scheduling change? Can it be changed? For example, if a contractor is casting a slab with SCC and schedules the traditional number of laborers for the placement crew, then this process has not been adjusted so as to take full advantage of the benefit. It may be that this understanding will develop over time as SCC is used more frequently.

Assess the Training Needs of the Organization

Training programs can be developed around the features, benefits and value, materials, proportions, and production, as well as around placing and curing or any other aspect of SCC. It is recommended that training programs within an organization be cross-functional. For example, a salesman for a ready-mixed concrete company looking to brand SCC should certainly go through the technical training. During training, people will learn the appropriate terminology allowing efficient communication both with the customer and with the internal technical department. The ability to effectively communicate is critical for implementation. Similarly, a concrete contractor should understand the value and benefits of SCC so as to make use of them in the best way possible. These cross-functional training programs more often than not serve as opportunities for further refinement of both economic and technical concepts. Training will be discussed in more detail later in this chapter.

Outline the Details of the Application and the Necessary SCC Performance

The details of the application include those reviewed in previous chapters, such as flowing distance and reinforcement levels. The outline should include both the fresh and hardened properties, along with the projected delivery and placement techniques. One should pay particular attention to those variables that may change throughout the course of a project, in particular the temperature and humidity changes that can occur from morning to afternoon or as the seasons change. By considering these variables beforehand, one creates a plan that then forms a large part of the training program.

Assess the Availability and Characteristics of the Local Raw Materials

A concrete producer, whether ready-mixed or precast, will know in general what material sources he or she has available. However, it is important to know the characteristics of these materials relative to the targeted concrete properties. This does not mean that the producer must obtain samples and conduct extensive testing on each and every material, because in many cases the material supplier will have the required data on hand or can generate it. These data can and should be cataloged for future reference. It is also helpful to know who, if anyone, is producing or using SCC in the area and to know what materials they are using, as long as they are willing to share that information. In some cases, a producer may be willing to discuss this information. Knowing the limitations of available materials is critical when developing an SCC mixture.

Create a Mixture Development and Mock-Up Plan

The size and extent of this testing plan depend on the level of SCC experience one has and whether the intention is to develop a project-specific mixture or investigate raw material influences. Either way, the plan should be clear and thought through before starting, and sufficient time should be set aside for its completion. Any test program targeting a specific project should consider the need for a mock-up test placement. The decision on this will depend on the project size and complexity, along with the contractor's and placement crew's experience and need for training. Once the plan has been created, it should be executed and adjusted as necessary.

The producer creates his quality plan at this point. The plan should provide direction for making mixture adjustments based on raw material characteristics or concrete testing results. It should also include the corrective action process in case of unintended results. How does the batch operator know that his mixture in the mixer is good-quality SCC? How often is SCC tested for its various fresh and hardened properties? What materials are tested and at what frequency? These, and questions like them, should be answered at this point.

Develop a Delivery and Placement Plan (Logistics)

The placement techniques have already been considered and decided upon, so this step is all about timing. In applications such as architectural elements, the placement of SCC is as important as the mixture properties. Ensuring that the concrete delivery rate is fast enough to satisfy the casting rate is important.

Training

A basic SCC training plan should answer three main questions: "What is it?", "Why should I use it?", and "How do I produce and/or use it?" Each individual or group involved in using SCC should have these questions answered for his or her particular role. The details involved and the intensity of training will vary between members, but ensuring that all receive adequate training will dramatically improve the likelihood of a successful implementation. The following members of the team should receive the most in-depth application training on SCC: concrete producer quality control manager and technicians, concrete batching personnel, and the contractor's or producer's casting foremen. In the author's opinion, one of the most important members of this team, in terms of ensuring consistent and successful SCC production, is the concrete batchman. This individual should receive extensive and detailed training on all aspects of SCC. The consistency of the batched concrete starts with this person.

A training program for SCC can be constructed in any number of ways, but all should include a mixture of classroom-style discussion and hands-on activities. The following is one example of how to construct a training program. This example program is intended for training those responsible for applying SCC technology who do not have significant prior experience with it. This program is divided into specific modules. Although some information must be repeated between modules, each one is a self-contained unit that is relevant to certain user groups, while combined they can provide an in-depth program on SCC technology. A title is given to each module, along with a brief explanation of its basic content. The details of each module can be created from information presented in earlier chapters of this book, other published documents, or personal experience.

1. Introduction, Terminology, and Benefits

The concept of SCC is introduced and defined. It should be clearly distinguished from high-slump concrete. This can be done by a full description of the potential benefits of SCC. At this point, a producer or contractor can also reveal the specific benefits they anticipate receiving through the use of SCC. This way, the entire team understands the implementation goal. Terminology that will be used throughout the other training modules is also presented at this time. Use of a common terminology improves the likelihood that performance needs and adjustments are communicated from placement crews to quality control personnel to sales people. Presenting examples of quantified benefits from a project or projects is more effective than presenting a list of possible savings.

Hands-On Exercise

Construct two small forms 600 mm by 1200 mm by 150 mm. Batch two mixtures: one an SCC mixture with 700 mm slump flow and the second a conventional slump mixture with a slump of 175 mm. Have the participants cast one form with the SCC mixture and the other with the conventional mixture and consolidate it with an internal vibrator. Use a stopwatch to record the placement time of each casting. The next day, strip the forms and examine the surface appearance. This exercise will provide the participants with an initial experience of the labor and time savings associated with SCC, as well as demonstrating its impact on surface finish.

Key Concepts

- Filling ability
- Fluidity
- Passing ability
- Static and dynamic stability
- Yield stress
- Plastic viscosity
- SCC benefits and savings
- Case studies
- Applications.

2. Fresh Properties, Testing, and Target Properties

There will be some repetition of terminology surrounding the fresh properties of SCC in this module. The goal should be to introduce the basic rheological concepts, the practical SCC properties, and how they relate to element characteristics and the desired benefits. This is also an appropriate time to discuss the interrelationship between properties such as fluidity and stability, and between test methods such as T50 and V-funnel measuring viscosity. A thorough description and demonstration of all relevant test methods should be given.

Hands-On Exercise

Make two SCC mixtures: one stable and the other unstable. Test each mixture using the relevant methods for fluidity, passing ability, and stability. This will give the participants the opportunity to see how each test method discriminates between mixtures with different characteristics.

Key Concepts

- The wide variety of SCC mixture types
- Interrelationship between properties
- Practical application of rheology.

3. Materials, Admixtures, and Proportioning

Take time to discuss each constituent material and the characteristics of each that influence fresh properties. Relate these back to the properties and test methods from the previous module. Explain the building blocks of concrete: paste, mortar, and finally concrete. Explain the changes to fresh properties as you move from paste to concrete. Note that in an SCC mixture there is a point where aggregates dominate and this transitions to paste domination as the aggregate content is decreased (or paste content is increased). Take the participants through the process of proportioning an SCC mixture on paper, then have them repeat the exercise for two or three different SCC mixtures with different raw materials. Have them batch these mixtures in laboratory-size batches and see the influence of proportions and materials.

Hands-On Exercise

The following is an effective exercise in demonstrating the importance of balancing the proportions of materials:

- Take the proportions of a successful SCC mixture.
- Have the participants create the paste fraction from this mixture, including admixtures. Mix sufficient paste so that it can be split into thirds, with each third still being of sufficient size.
- Take two-thirds of the paste and add the appropriate volume of fine aggregate to it based on the mixture proportions. The participants should be directed to evaluate the change in flowing properties between the paste

(one-third is still left) and the just-created mortar. Split this mortar in half.

- Take one-half of the mortar and add the appropriate volume of coarse aggregate to it, again based on the mixture proportions.
- The participants can now compare the flowing properties of the paste, mortar, and SCC mixture.
- Each participant can now add more water, chemical admixtures, and fine or coarse aggregates to the paste, mortar, or concrete and witness the changing fresh properties. It is recommended that small changes be made at first to simulate batching errors, etc.

Key Concepts

- Powder composition
- Water demand
- Particle shape
- Particle size distribution
- Paste volume
- Water/powder ratio by volume (wv/pv)
- Void content
- Water content
- Main proportioning variables that influence rheological properties
- Robustness.

4. Production and Delivery

The purpose of this module is to raise awareness of the production issues that influence SCC properties. This includes everything from raw material storage, monitoring, and batching sequence to mixer types. For the concrete producer, it may be appropriate to request specific and detailed training from the batching software representative. *Training the batching, mixing, and quality control team to be proficient in using all of the monitoring and analysis tools available within the software should be a goal.* Each concrete plant is different, and therefore a production and delivery plan should be created for each. Because of its faster placement rate, the delivery of SCC may require a different logistics plan than conventional concrete. Fresh properties, such as workability retention, should also be considered when the delivery plan is created. Mixer truck drivers also need to be trained to ensure that all water is out of their drums before batching and that no extra water is ever added. If fluidity must be increased on site with SCC, this should be done with an HRWR admixture.

Hands-On Exercise

Hands on batching and mixing through a production plant is recommended to allow the participants to witness the influence of mixing time and to use the mixer monitoring equipment to judge the acceptance of a batch. It is also effective to batch a mixture, remove a sample for testing, and add water or other materials to the batch to show the influence on the mixer monitoring equipment and on the fresh properties.

Key Concepts

- Stockpiling raw materials and monitoring their properties
- Controlling moisture contents
- Aggregate and mixer moisture meters
- Amp meter
- Quality control testing
- Batching sequence
- Delivery logistics
- Placement rate.

5. Placement and Curing

The placement technique can influence the final results of an SCC project. The equipment used, along with the required flowing distance, drop height, and flowing and placement rates, affect the surface finish and can influence how a mixture is proportioned. Air can be entrapped in the bulk concrete as it is discharged from the mixer to the delivery vessel and as it is being placed into the form. Care should be taken to consider placement methods before the start of the pour. Proper moist curing should be done with SCC. Adequate moist curing becomes more critical as the wv/pv ratio decreases, as this will reduce the mixture's bleeding characteristics.

Hands-On Exercise

Take two jars, and pour water into one of them, allowing it to fall from a height of 150 mm above the jar. Each participant should note the air bubbles being pulled into the water and rising to the top and escaping. Into the second jar, pour, in the same way, a more viscous, translucent material, such as corn syrup or motor oil. Each participant should note the bubbles being pulled into the more viscous material and not escaping as quickly. This is intended to provide insight into the interaction between placement technique, mixture properties, and surface finish.

Key Concepts

- Placement techniques
- Continuous versus discontinuous placement
- Placement flow direction versus filling flow direction
- Placement turbulence
- Box-outs
- Horizontal versus vertical elements
- Formwork pressure
- Bleeding
- Moist curing.

6. Hardened Properties and Specifications

When comparing SCC with a conventional concrete mixture, general statements about relative performance should be avoided. Broad-brushed statements can be misleading. Like those of conventional concrete mixtures, the hardened properties of SCC are primarily governed by the mixture proportions and follow the same rules of thumb. If shrinkage, creep, modulus of elasticity, or other properties are important, then those targets should be included and addressed during the mixture proportioning process. Some properties, such as bond to prestressing strand, are still being tested. SCC can be proportioned to achieve a wide range of hardened properties and cannot be specified through a prescriptive method. The desired performance characteristics should be specified and the producer should be responsible for delivering that performance.

Key Concepts

- Wide range of properties possible
- Mixture proportions
- Paste volume
- Water content
- Sand-to-aggregate ratio
- Water/cement (w/c) ratio
- Prescriptive versus performance specification.

Summary

Planning and preparation are important keys to implementing SCC. The process should follow a systematic approach, and a general outline has been presented here. The details of each step, however, will contain nuances specific to the particular project or facility. Those interested in using SCC should invest the time to plan properly. This includes the concrete producer as well as the contractor. Cross-functional training is a critical part of the implementation plan and all members of a team should receive training. Adequate training is critical for quality control, batching, and placement personnel.

Chapter 15

Applications of SCC

Introduction

In the more than 20 years since it was first developed, SCC has been used with tremendous success on numerous projects with widely varying requirements throughout North America and the world. Some of these projects have been reported in published case studies, details of several of which have been presented in the preceding chapters of this book. This chapter, however, is intended to provide a review of the variety of ways in which SCC has and can be used. The projects highlighted have been chosen to demonstrate this diversity of application, which includes precast, cast-in-place, structural, architectural, vertical, horizontal, large, and small projects. The projects have also been chosen to demonstrate the variety of ways in which SCC can be and is placed, such as by pumping, crane and bucket, mixer truck, and others. The real value of this chapter is that the practitioner, especially one who has never before used SCC, can clearly see the results of successful SCC projects.

Barracks Reconstruction at the Citadel Military College in Charleston, South Carolina

Two student housing barracks at the Citadel Military College were reconstructed, with both projects using significant volumes of SCC. The walls and columns of each structure are exposed to view after receiving only a light coat of paint; therefore, the final as-cast surface finish was very important. Placement and final finish problems had been experienced on similar projects previously completed on campus. These difficulties resulted in the need for surface patching, and ultimately lead to higher costs.[1]

The Padgett-Thomas Barracks was the first to be rebuilt. The original concrete specification called for a 100 mm slump, which, after early discussion, was adjusted to 175 mm with the use of high-range water reducer (HRWR).[1] The adjusted mixture did not provide the desired surface finish, with significant rock pockets and surface blemishes occurring. An SCC mixture with a slump flow of 675 mm was developed by the concrete producer and admixture supplier and was pumped into place. Placement times were reduced by 50% and the number of laborers was decreased from six to two. The resulting surface finish was of high quality, requiring little or no patching, and all corners and edges were very sharp. Figure 15.1 shows the surface finish of the SCC versus conventional concrete over a doorway. The left side is SCC and the right is the conventional concrete.

After stripping the forms, the walls were essentially ready for painting. More than 5000 m³ of SCC were used on this structure. Figure 15.2 shows the sharp edges and blemish-free as-cast surface finish obtained on the project.

The Padgett-Thomas Barracks re-opened in 2004. That same year, the Law Barracks was closed and then demolished in 2005. The same contractor, Ellis Don Construction, and concrete supplier, Ready-Mixed Concrete, were selected for the job. Because of the success on the first project, SCC was written into the concrete specifications. Over 2800 m³ of SCC were used, and the project was completed ahead of schedule, with excellent results. The Law Barracks reconstruction was chosen by *Concrete Construction Magazine* as one of the top projects for 2005 for its study of SCC proportioning and

Figure 15.1 SCC on Left versus Conventional Concrete on Right. (Courtesy of Richard Morrow.)

Figure 15.2 Intricate Formwork and Crisp Lines. (Courtesy of Richard Morrow.)

performance, in particular formwork pressures that were evaluated throughout the duration of the project.[2]

National Aeronautical and Space Administration (NASA) Vibro-Acoustic Test Facility at Plum Brook Station

Walls for the world's largest environmental simulation chamber designed to test hardware for the United States' Orion and Constellation space programs were cast with SCC.[3] As part of the renovated test facility, a vibro-acoustic testing chamber was constructed. This chamber contains 24 horns that will blow high-pressure nitrogen gas to match the intensity level of the sounds produced during spacecraft launch. The locations for all 24 horns were installed into the wall formwork before concrete placement. As can be seen in Figure 15.3, these box-outs created a maze through which the SCC was required to flow. An SCC mixture with a 650 mm slump flow was chosen because of its ability to flow around these box-outs and through reinforcement while providing a smooth, blemish-free surface. Concrete placement occurred via a concrete pump, and the final results were excellent.

Figure 15.4 shows the smooth surface of the wall opposite the horn wall.

The Oceans Exhibit at the Indianapolis Zoo

The architectural walls for a saltwater shark tank were cast with SCC.[4] To satisfy the aesthetic requirements of the project, the mixture was developed

Figure 15.3 Box-Outs in Horn Wall. (Courtesy of Duane Rose.)

Figure 15.4 Opposite Wall. (Courtesy of Duane Rose.)

using white cement and integrally mixed color. During construction, the SCC was pumped into place and flowed under the window wells, through reinforcement, and around the form-liner details. The result was a blemish-free surface, with no voids under the window wells and sharp lines along the reveals. Figures 15.5 and 15.6 show the details of the tank walls.

Precast Utility Vaults

SCC accounts for a significant percentage of the mixture types used in precast production in North America. The typical precast producer both

Figure 15.5 Detailed Reveals in Shark Tank Wall.

Figure 15.6 Reveals and Window Well of Shark Tank Wall.

mixes and places concrete, thereby internally collecting many of the benefits associated with SCC use. The utility vaults shown in Figures 15.7 and 15.8 were cast with a 650–700 mm slump flow SCC mixture with a T50 time of approximately 1–1.5 seconds. A single laborer, controlling the crane and bucket, was involved in the casting of these elements. Notice in the figures the detailed edges and the obstacles around which the mixture was required to flow.

Prestressed Double-Tees

SCC is used in casting prestressed double-tees in many locations across North America. Figure 15.9 shows a 500–550 mm slump flow mixture being cast into the stem of a double-tee. The SCC at this precast plant is delivered and placed by mixer truck. The standard casting procedure is to fill both stems, then cast the deck. In this case, SCC is used primarily to reduce placement time and labor and to improve the final surface finish in order to improve quality and reduce patching labor requirements. Figure 15.10 shows the as-cast surface finish of the stem and the underside of the deck. This highlights the ability of lower-slump-flow SCC mixtures with a balanced viscosity to produce an excellent surface finish.

Figure 15.7 Precast Utility Vaults Cast with SCC.

Figure 15.8 Obstacles around which the SCC Was Required to Flow.

Figure 15.9 Casting the Stem of a Double-Tee.

Figure 15.10 As-Cast Surface Finish of Double-Tee Stem and Underside of the Deck.

City of Cleveland Municipal Waste Water Treatment Plant

Underground sedimentation tanks were being rehabilitated, and the new concrete floors for these tanks were cast with SCC. The concrete was pumped from above ground several hundred meters through pump lines running into the underground tanks (Figures 15.11 and 15.12). The SCC had a slump flow of 500–600 mm and received a hard troweled surface finish. Figure 15.13 shows the placement process.

Figure 15.11 Above-Ground SCC Delivery and Pumping.

Figure 15.12 Underground Pump
Lines for SCC Placement.

Figure 15.13 Placement Process for SCC Used to Cast New Tank Floors.

Reconstruction of the I-35 W Bridge, Minneapolis, Minnesota

SCC was used to cast the drilled shafts for the I-35 W bridge, some of which were 2.4 m in diameter. The mixture had a slump flow of approximately 700 mm and was required to flow through congested reinforcement. In addition, because these elements were considered mass concrete, they were proportioned with 60% of the cementitious material being pozzolans.[5] In addition to this project, SCC has been used in drilled shafts in South Carolina, New Jersey, and Virginia. It has also been evaluated for use on projects in Hawaii and Alabama.[6]

Laboratory Sedimentation Pit Construction

During the renovation of several laboratories at the BASF Technical Center in Cleveland, Ohio, new sedimentation pits were constructed. These areas serve as one of the primary locations for research and development on new construction materials. The construction of the pits required excavation of the holes and then installation of the formwork. SCC was chosen because of the need to flow around obstacles, including the box-out windows in the center of the pits. The mixture had a slump flow of approximately 650 mm and was placed by wheelbarrow. Because of the slower placement method, the SCC was developed to maintain its workability for a sufficient time to ensure that no pour lines or cold joints occurred. Figures 15.14 and 15.15 show the sedimentation pit details after casting.

Figure 15.14 Sedimentation Pit.

Figure 15.15
Underside of
Box-Out and
Surface Finish
on Walls.

The Cathedral of Christ the Light, Oakland, California

The curved concrete reliquary walls that function structurally and as exposed architectural surfaces were cast with SCC.[7] SCC was chosen for numerous reasons, including congested reinforcement in the walls and the need for sharp details and a uniform appearance on the exposed walls, among others. Before applying SCC in the casting of these critical, architectural walls, the contractor used the mixture to cast the underground mausoleum walls in an effort to become familiar with the technology. The mixture had a slump flow of 610–690 mm and was placed by pumping. Figures 15.16–15.18 show the wall details of the completed cathedral.

Figure 15.16 Cathedral Walls. (Courtesy of Skidmore Owings & Merrill, Photographer Cesar Rubio.)

Figure 15.17 Details of the Exposed Concrete Walls Cast with SCC. (Courtesy of Skidmore Owings & Merrill, Photographer Timothy Hursley.)

Figure 15.18
Exposed Cathedral Walls Cast with SCC. (Courtesy of Skidmore Owings & Merrill, Photographer Timothy Hursley.)

Summary

As can be seen from the examples presented in this chapter, SCC can be used in almost any kind of application, whether it is large or small, horizontal flatwork, vertical columns, architectural, structural, precast, or cast-in-place. Additionally, the placement techniques used on these projects are as varied as for conventional concrete. SCC does require some changes to materials, methodologies, and techniques, but its potential benefits are indisputable.

Concluding Remarks

The intention of this book has been to build a link between the worlds of materials science/academia and construction practice with respect to self-consolidating concrete. A great deal of research on the properties of SCC has been conducted over the last 20 years. This research has led to the development of numerous test methods, mixture proportioning techniques, materials, and chemical admixtures. It has renewed the industry's focus on aggregate characteristics, the use of waste material for fines, placement techniques, and formwork pressure calculations. SCC has expanded our interest in rheological measurements and their meaning in practice and it has led to the development of new production monitoring equipment. In short, the advent of SCC has initiated a review of many aspects of concrete materials and construction. At the same time, SCC has been and is being used to produce concrete elements and structures every day. Millions of cubic meters of SCC have been placed since its initial development in Japan and it certainly appears that this trend will continue.

Although there is always room for more research and a finer understanding of SCC's properties, the area of enhanced production control and production/placement techniques can and will benefit from further investigative efforts. These efforts would likely result in an even broader acceptance and use of SCC. For example, in a 2011 survey given to the American Society of Concrete Contractors (ASCC), when asked the question "Which one statement best describes your awareness of SCC?", 40% of respondents said they had used it successfully and planned to use it again, less than 5% said they had used it and would not use it again, and 50% said they had heard of it, could see its potential value, but had never tried it. This was a clear indication that SCC has the potential to bring real value to the concrete contractor. However, in the same survey, some contractors gave comments such as "The producer really needs to be on top of his game," and "Some producers can deliver it and some cannot." A follow-up discussion with contractors indicated that they look for the producer that can consistently deliver quality SCC and this is when the value is truly present. It is because of these comments, as well as comments in personal discussions with members of the precast industry regarding control of mixture stability, that the need for focusing on production processes has become more evident. This is not to say that consistent SCC is not being produced, because it most certainly is by some very good and conscientious producers. However, the above comments suggest that room does exist among the overall population of producers to improve on the consistency of SCC production.

One common element that seems to exist among those producers successfully producing SCC is a highly qualified quality control department and batch controller. In general, these individuals are well versed in the quality and production control software packages that their firms use. They know how to move in and out of the various programs and regularly spend time reviewing past materials and mixture performance data in an effort to further refine their ability to adjust individual production batches and mixture proportions. Educating the batch controller and the quality control department is an essential part of the process. Additionally, although the amount of quality effort tends to be higher during the initial introduction of SCC, this equilibrates to a lower level once the internal expertise for batching and controlling is developed.

The contractor or precast placement crew should expect to modify their practices and realize that there will be a learning curve when applying or using SCC for the first time. Every new tool has a learning curve—even a young person goes through a learning curve when pounding nails with a hammer for the first time. In many cases, mock-up elements serve to bring both the concrete producer and contractor together on the same page regarding the application of SCC. On some projects, the contractor has chosen to employ SCC on the less critical sections, using this as an opportunity for developing practical experience. Once a contractor or producer has achieved success, more successful applications tend to follow.

A review of the previous chapter, along with the numerous case studies published, clearly demonstrates SCC's usefulness to the industry. SCC is an exciting tool with a broad history of success around the world. The benefits it provides are available to anyone with the desire to take advantage of them. Hopefully, this book can provide some assistance in turning that desire into reality.

Glossary

Admixture A material other than water, aggregates, cementitious materials, and fiber reinforcement, used as an ingredient of a cementitious mixture to modify its freshly mixed, setting, or hardened properties and that is added to the batch before or during its mixing (ACI terminology).

Amp meter A device for measuring the current in amps passing through it used to gauge the workability level of concrete in a concrete mixer.

Bleeding The autogenous flow of mixing water within, or its emergence from, a newly placed cementitious mixture caused by the settlement of solid materials within the mass (ACI terminology).

Blocking The condition occurring when pieces of aggregate come into contact and do not pass one another, thereby stopping flow of the concrete in that area. This can happen more often as the space within which the concrete is flowing begins to restrict, such as in detailed formwork, through tightly spaced reinforcement, or even through a pump line reducer.

Box-out An area formed into formwork designed to create a void in the final cast element such as a window or doorway.

Bridging *See* Blocking.

Bugholes, also known as **surface air voids** Small regular or irregular cavities, usually not exceeding 15 mm in diameter, resulting from entrapment of air bubbles in the surface of formed concrete during placement and consolidation (ACI terminology).

Column segregation A test method used to determine the segregation resistance of SCC (ASTM C 1610).

Consistency The degree to which a freshly mixed concrete, mortar, grout, or cement paste resists deformation (ACI terminology).

Conventional concrete Concrete with a conventional workability measured by slump.

Dispersant A material that deflocculates or disperses finely ground materials by satisfying the surface energy requirements of the particles (ACI terminology).

Dynamic stability The characteristic of SCC that determines its ability to maintain a homogenous distribution of its constituent materials during transport and placement.

Fillers Fine material other than supplementary cementitious materials (SCMs) used to increase the powder content of SCC.

Filling ability The ability of the fresh concrete mixture to flow into and fill formwork under its own weight. This term can be used interchangeably with **fluidity** and **flowability** as they pertain to SCC.

Fineness A measure of particle size (ACI terminology).

Flowability *See* Filling ability.

Fluidity *See* Filling ability.

Form filling flow direction The direction in which the concrete will flow once inside the form, when filling an empty form with SCC.

High-range water reducer (HRWR), also known as a **high-range water-reducing admixture** or **superplasticizer** A water-reducing admixture capable of producing large water reduction or great flowability without causing undue set retardation or entrainment of air in mortar or concrete (ACI terminology).

High-range water-reducing admixture *See* High-range water reducer.

Instability The tendency for the constituent materials of an SCC mixture to separate.

J-ring A test method to measure the passing ability of SCC.

L-box A test method to measure the passing ability of SCC.

Mortar A mixture of cement paste and fine aggregate (ACI terminology).

Mortar halo A noticeable concentration of paste or mortar that does not contain coarse aggregate, formed at the perimeter of the slump flow patty.

Particle shape The form of a particle (ACI terminology): rounded, angular, cubical, elongated, flat.

Particle size distribution Grading.

Passing ability The ability of an SCC mixture to flow through restricted spaces without blocking.

Paste Binder of concrete and mortar consisting essentially of cement, water, hydration products, and any admixtures, together with very finely divided materials included in the aggregates (ACI terminology); air content is considered as a component of paste.

Patching *See* Rubbing.

Penetration apparatus The test equipment used for ASTM C 1712.

Placement flow direction The direction of concrete flow as SCC is being placed into an empty form.

Pour lines Visible lines on vertically cast surfaces that clearly distinguish separate deliveries of concrete.

Powder Material with a particle size passing the No. 100 sieve, or close to 125 μm or smaller intentionally added to an SCC mixture for the intention of increasing paste volume and/or impacting the concrete rheology.

Retemper To add water and remix a cementitious mixture to restore workability to a condition in which the mixture is placeable or usable (ACI terminology).

Rheology   The science dealing with the deformation or flow of materials.

Rheometer, also known as a viscometer   A device used to measure rheological properties, in this case, the properties of SCC.

Robustness   The insensitivity of an SCC mixture to fluctuations in batching accuracy and/or raw material characteristics such as moisture contents and particle size distribution that could lead to changes in the mixture's fresh properties.

Rubbing, also known as patching or sack rub   A finish for formed concrete surfaces, designed to produce even texture and fill pits and air holes; after dampening the surface, mortar is rubbed over the surface; then, before the surface dries, a mixture of dry cement and sand is rubbed over it with either a wad of burlap or a sponge-rubber float to remove surplus mortar and fill voids (ACI terminology).

Sack rub   *See* Rubbing.

SCC   Self-compacting or self-consolidating concrete.

Segregation resistance   The ability of SCC to resist separation of its constituent materials.

Self-compacting concrete   *See* Self-consolidating concrete.

Self-consolidating concrete, also known as self-compacting concrete (SCC)   A highly flowable, non-segregating concrete that can spread into place, fill the formwork, and encapsulate the reinforcement without any mechanical consolidation (ACI 237).

Sieve segregation   Test for measuring the segregation resistance of SCC mixtures.

Slump flow   A test method used to measure the fluidity or filling ability of SCC (ASTM C 1611), in which the lateral spread of the resulting concrete paddy is measured after removal of a slump cone.

Spacing factor   An index related to the maximum distance of any point in a cement paste or in the cement paste fraction of mortar or concrete from the periphery of an air void; also known as Powers' spacing factor (ACI terminology).

Spread   *See* Slump flow.

Stability   *See* Segregation resistance.

Static stability   The characteristic of SCC that determines its ability to maintain a homogenous distribution of its constituent materials after placement until the onset of setting.

Superplasticizer   *See* High-range water reducer.

Supplementary cementitious material (SCM)   Inorganic material such as fly ash, silica fume, metakaolin, or ground-granulated blast-furnace slag (GGBFS) that reacts pozzolanically or hydraulically (ACI terminology).

Surface air voids   *See* Bugholes.

T50 time   The time required for a slump flow to reach a diameter of 50 cm, also known as T500 and T20 for millimeters and inches, respectively.

Thixotropy   The ability of a material to change from stiff to fluid to stiff again when agitation or shear is applied and then removed.

U-box   A test method developed in Japan to measure the passing ability of SCC.

V-funnel   A test method developed to measure the efflux time of a set volume of concrete, which can be correlated to concrete viscosity.

Viscometer   *See* Rheometer.

Viscosity   A material's internal resistance to flow. This rheological term can be qualitatively understood as a mixture's degree of cohesiveness or stickiness.

Viscosity-modifying admixture (VMA)   A separately added admixture that when included in concrete changes the viscosity of the mixture.

Visual stability index (VSI)   The qualitative assessment of the overall stability of an SCC mixture. This is typically determined by evaluating a slump flow patty for a mortar halo or aggregate pile and giving a rating of 0, 1, 2, or 3 to the mixture.

Void content (of aggregate)   The volume of void space between aggregate particles that must be filled with paste.

Workability   The property of freshly mixed concrete that determines the ease with which it can be mixed, placed, consolidated, and finished.

Workability-retaining admixture   An admixture that allows concrete or mortar to maintain its fresh characteristics throughout the transportation, placement, consolidation, and finishing operations without adversely affecting the time of setting or the hardened properties of the concrete or mortar.

Workability retention   The ability of a concrete mixture to maintain a level of workability over a certain period of time.

wv/pv ratio   The ratio of the volume of water to the volume of total powder in an SCC mixture.

Yield stress   A rheological property that is the stress required to initiate flow in a material.

References

Chapter 1 Introduction, History, and Acceptance of SCC

1. Skarendahl, A., "SCC is an Important Step Towards Industrialization of the Building Industry," *Proceedings of the First International Symposium on Self-Compacting Concrete, Stockholm, Sweden*, 1999, pp. 15–21.
2. "RILEM Technical Committee FPC, Form Pressure Generated by Fresh Concrete," March 4, 2011, www.RILEM.net.
3. "Well-Equipped MBO Precast Sets NPCA Plant Certification Record," *Concrete Products Magazine*, December 2008, p. PC6.
4. Hackley, V.A., Ferraris, C.F., *Guide to Rheological Nomenclature: Measurements in Ceramic Particulate Systems*, NIST Special Publication 946, National Institute of Standards and Technology, 2001, p. 9.
5. Tattersall, G.H., Banfill, P.F.G., *The Rheology of Fresh Concrete*, Pitman, Boston, 1983.
6. Okamura, H., Ouchi, M., "Self-Compacting Concrete, Development, Present Use and Future," *Proceedings of the First International Symposium on Self-Compacting Concrete, Stockholm, Sweden*, 1999, p. 4.
7. *European Guidelines for Self-Compacting Concrete: Specification, Production and Use*, European Project Group, 2005.
8. "ACI 237R-07—Self-Consolidating Concrete," *ACI Manual of Concrete Practice*, American Concrete Institute, 2007, p. 2.
9. Jacobs, F., Hunkeler, F., "Design of Self-Compacting Concrete for Durable Concrete Structures," *Proceedings of the First International Symposium on Self-Compacting Concrete, Stockholm, Sweden*, 1999, p. 399.
10. "ACI 302.1R-04, Guide for Concrete Floor and Slab Construction," *ACI Manual of Concrete Practice*, American Concrete Institute, 2010, p. 33.
11. "ACI 363.2R-98, Guide to Quality Control and Testing of High Strength Concrete," *ACI Manual of Concrete Practice*, American Concrete Institute, 2010, p. 4.
12. Khayat, K., "Opening Remarks, SCC 2010 Conference," Presented in Montreal, Canada, September 27, 2010.
13. Ozawa, K., Maekawa, K., Kunishima, M., Okamura, H., "Development of High Performance Concrete Based on the Durability Design of Concrete Structures," *Proceedings of the Second East-Asia and Pacific Conference on Structural Engineering and Construction*, 1989, Vol. 1., pp. 445–450.
14. Rogers, E.M., *Diffusion of Innovations*, 5th ed., Free Press, New York, 2003.
15. Daczko, J., "North American Acceptance of Self-Consolidating: A Diffusion of Innovations Perspective," *Concrete Plant International*, April 2009, pp. 18–21.
16. Frank, D., "Acceptance of SCC—Precast Concrete Industry Perspective," Presented at the Fall 2008 ACI Convention, St Louis, November 4, 2008.

17. Lobo, C., "Acceptance of SCC—Ready-Mixed Perspective," Presented at the Fall 2008 ACI Convention, St Louis, November 4, 2008.
18. BASF, "Concrete Technology for Contractors Survey," March 2011.

Chapter 2 Benefits and Limitations of SCC

1. *European Guidelines for Self-Compacting Concrete: Specification, Production and Use,* European Project Group, 2005.
2. "ACI 237R-07—Self-Consolidating Concrete," *ACI Manual of Concrete Practice,* American Concrete Institute, 2007, pp. 2–8.
3. Skarendahl, A., Billberg, P., *Report 35, Casting of Self-Compacting Concrete, Final Report of RILEM Technical Committee 188-CSC: Casting of Self-Compacting Concrete,* RILEM Publications, 2006.
4. Ready-Mixed Concrete Association of Ontario (RMCAO), "SCC Calculator, December 2009," www.rmcao.org.
5. www.selfconsolidatingconcrete.org/benefits, March 2011.
6. Mizobuchi, T., Yanai, S., Takada, K., Sakata, N., Nobuta, Y., "Field Applications of Self-Compacting Concrete with Advantageous Performances," *Proceedings of the First International RILEM Symposium on Self-Compacting Concrete, Stockholm, Sweden,* 1999, pp. 605–616.
7. Nishizaki, T., Kamada, F., Chikamatsu, R., Kawashima, H., "Application of High Strength Self-Compacting Concrete to Prestressed Concrete Outer Tank for LNG Storage," *Proceedings of the First International RILEM Symposium on Self-Compacting Concrete, Stockholm, Sweden,* 1999, pp. 629–638.
8. Billberg, P., Petersson, O., Osterberg, T., "Full Scale Casting of Bridges with Self-Compacting Concrete," *Proceedings of the First International RILEM Symposium on Self-Compacting Concrete, Stockholm, Sweden,* 1999, pp. 639–650.
9. Chikamatsu, R., Shinkai, C., Kushigemachi, H., "Application of Low Shrinkage Type Self-Compacting Concrete to an Advanced Large Water Purification Plant," *Proceedings of the First International RILEM Symposium on Self-Compacting Concrete, Stockholm, Sweden,* 1999, pp. 659–668.
10. Pichler, R., "The Use of SCC for Building the Millennium Tower in Vienna," *Proceedings of the First International RILEM Symposium on Self-Compacting Concrete, Stockholm, Sweden,* 1999, pp. 729–731.
11. Inoue, H., Takeichi, Y., Ohtom, T., "Construction of Rigid Foundation of Underground Diaphragm Walls with Highly Congested Reinforcing Bar Arrangements by Using Self-Compacting Concrete," *Proceedings of the Second International RILEM Symposium on Self-Compacting Concrete, Tokyo, Japan,* 2001, pp. 643–650.
12. Tanaka, M., Mori, K., Shindoh, T., Sakamoto, J., "Application of Self-Compacting Concrete to Steel Segments of Multi-micro Shield Tunneling Method," *Proceedings of the Second International RILEM Symposium on Self-Compacting Concrete, Tokyo, Japan,* 2001, pp. 651–660.
13. Ohtomo, T., Asaka, S., Kim, J., Park, C., Beak., S., Jung, C., Tsutsumi, Y., "Self-Compacting Concrete Used for Underground Diaphram Walls of the World's Largest 200,000 kl In-ground Tanks at Inchon LNG Terminal in Korea," *Proceedings of the Second International RILEM Symposium on Self-Compacting Concrete, Tokyo, Japan,* 2001, pp. 661–670.
14. Daczko, J., Phillips, S., "Self-Compacting Concrete in Underground and Mining Applications," *Proceedings of the Second International RILEM Symposium on Self-Compacting Concrete, Tokyo, Japan,* 2001, pp. 671–680.

15. Johansen, K., Kyltveit, B., "SCC in a Rock Repository for Radioactive Waste," *Proceedings of the Second International RILEM Symposium on Self-Compacting Concrete, Tokyo, Japan*, 2001, pp. 681–686.

16. Walraven, J., "Self-Compacting Concrete in the Netherlands," *Proceedings of the First North American Conference on the Design and Use of Self-Consolidating Concrete, Chicago, Illinois*, 2002, pp. 355–360.

17. Lessard, M., Talbot, C., Phelan, W.S., Baker, D., "Self-Consolidating Concrete Solves Challenging Placement Problems at the Pearson International Airport in Toronto, Canada," *Proceedings of the First North American Conference on the Design and Use of Self-Consolidating Concrete, Chicago, Illinois*, 2002, pp. 367–370.

18. Faornasier, G., Giovambattista, P., Zitzer, L., "Self-Consolidating Concrete in Argentina: Development Program and Applications," *Proceedings of the First North American Conference on the Design and Use of Self-Consolidating Concrete, Chicago, Illinois*, 2002, pp. 389–394.

19. Ramsburg, P., Ludirdja, D., Masik, O., "Utilizing SCC in Double Cycling Prodcution of Double Tee Beams," *Proceedings of the Second North American Conference on the Design and Use of Self-Consolidating Concrete and the Fourth International RILEM Symposium on Self-Compacting Concrete, Chicago, Illinois*, 2005, pp. 1049–1054.

20. Benedict, D.E., Worsfold, S.J., "Economically Attractive Field Applications for Self-Consolidating Concrete," *Proceedings of the Second North American Conference on the Design and Use of Self-Consolidating Concrete and the Fourth International RILEM Symposium on Self-Compacting Concrete, Chicago, Illinois*, 2005, pp. 1055–1060.

21. Chai, P.H.-W., Yang, M., "Case Study: RC Building Rehabilitated with Self-Compacting Concrete," *Proceedings of the Second North American Conference on the Design and Use of Self-Consolidating Concrete and the Fourth International RILEM Symposium on Self-Compacting Concrete, Chicago, Illinois*, 2005, pp. 1069–1074.

22. Fernandez, P., Luciano, J., Constantiner, D., "Successful Implementation of SCC in a Precast Operation—A Case Study," *Proceedings of the Second North American Conference on the Design and Use of Self-Consolidating Concrete and the Fourth International RILEM Symposium on Self-Compacting Concrete, Chicago, Illinois*, 2005, pp. 1075–1082.

23. Jooste, P., Fanourakis, G., "SCC: The South African Experience," *Proceedings of the Second North American Conference on the Design and Use of Self-Consolidating Concrete and the Fourth International RILEM Symposium on Self-Compacting Concrete, Chicago, Illinois*, 2005, pp. 1119–1124.

24. Vieira M., Bettencourt, A., "Repair Applications of Self-Compacting Concrete," *Proceedings of the Second North American Conference on the Design and Use of Self-Consolidating Concrete and the Fourth International RILEM Symposium on Self-Compacting Concrete, Chicago, Illinois*, 2005, pp. 1125–1132.

25. Khrapko, M., "Experience with Supplying SCC," *Proceedings of the Second North American Conference on the Design and Use of Self-Consolidating Concrete and the Fourth International RILEM Symposium on Self-Compacting Concrete, Chicago, Illinois*, 2005, pp. 1189–1194.

26. Collepardi, M., Collepardi, S., Ogoumah Olagot, J.J., Troli, R., "Laboratory Tests and Field Experiences of High Performance SCCs," *Proceedings of the Third International RILEM Symposium on Self-Compacting Concrete, Reykjavik, Iceland*, 2003, pp. 904–912.

27. Khayat, K.H., Morin, R., "Performance of Self-Consolidating Concrete used to Repair Parapet Wall in Montreal," *Proceedings of the Third International*

RILEM Symposium on Self-Compacting Concrete, Reykjavik, Iceland, 2003, pp. 913–919.

28. Sciaraffia, R., "SCC Application in a Tunnel Lining (Chile)," *Proceedings of the Third International RILEM Symposium on Self-Compacting Concrete, Reykjavik, Iceland,* 2003, pp. 986–994.

29. Utsi, S., Jonasson, J., Wallin, K., Ekman, T., "Use of SCC in a Tunnel Lining for a Railway Tunnel in Sweden," *Proceedings of the Third International RILEM Symposium on Self-Compacting Concrete, Reykjavik, Iceland,* 2003, pp. 995–1004.

30. Paris, M., Tavano, S., Albiero, E., "Self-Compacting Concrete (SCC): The Construction of the Base for the Mill of the Unimed Cement Plant of Barletta (Italy)," *Proceedings of the Third International RILEM Symposium on Self-Compacting Concrete, Reykjavik, Iceland,* 2003, pp. 1017–1025.

31. Titherington, M., Schell, H., "Use of Self-Consolidating Concrete for Bridge Repairs," *Proceedings of the Third North American Conference on the Design and Use of Self-Consolidating Concrete, Chicago, Illinois,* 2008.

32. Goldman, A., Sokal, J., "SCC as Mass Concrete with Blast Furnace Slag Cement," *Proceedings of the Third North American Conference on the Design and Use of Self-Consolidating Concrete, Chicago, Illinois,* 2008.

33. Talbot, C., "Behavior of Self-Consolidating Concrete at Cryogenic Temperatures," *Proceedings of the Third North American Conference on the Design and Use of Self-Consolidating Concrete, Chicago, Illinois,* 2008.

34. Peterson, E., "Meeting the Formwork Requirements for a Challenging Structure," *Concrete International,* Vol. 32. No. 1, January 2010, pp. 42–48.

35. Rodriguez Camacho, R.E., Afif, U., Martinez Corona, G., Montano Ramon, H., Martinez Sanchez, B., "Applications of SCC Technology for Precast/Prestressed Elements in Mexico," *Concrete Plant International,* June 2007, pp. 62–70.

36. Abbuhl, J., Welch, M., "Application of Architectural Self-Consolidating Concrete," *Proceedings of SCC2010, Design, Production and Placement of Self-Consolidating Concrete, Montreal, Canada,* 2010, pp. 1355–1358.

37. Yanbin, T., Huajian, L., Yongjiang, X., Zhonglai, Y., "Study on Property of SCC used in High Speed Railway Slab Ballastless Track," *Proceedings of SCC2010, Design, Production and Placement of Self-Consolidating Concrete, Montreal, Canada,* 2010, pp. 1359–1368.

38. Galeota, D., Giammatteo, M., Gregori, A., Manfredi, G., Marino, R., Vivaldi, S., "SCC for Rapid Home Construction after a Natural Disaster," *Proceedings of SCC2010, Design, Production and Placement of Self-Consolidating Concrete, Montreal, Canada,* 2010, pp. 1369–1377.

39. McSaveney, L., Khrapko, M., Papworth, F., "Self-Compacting Concrete for Superior Marine Durability: New Zealand's New Tauranga Harbour Link," *Proceedings of SCC2010, Design, Production and Placement of Self-Consolidating Concrete, Montreal, Canada,* 2010, pp. 1387–1396.

40. Schwartzentruber, L., Lenglet, C., Larive, C. "Innovative Design and Placement of Frost Resistant Self-Compacting Concrete to Repair Tunnel Lining," *Proceedings of SCC2010, Design, Production and Placement of Self-Consolidating Concrete, Montreal, Canada,* 2010, pp. 1397–1405.

41. Martin, D., "Economic Impact of SCC in Precast Applications," *Proceedings of the First North American Conference on the Design and Use of Self-Consolidating Concrete, Chicago, Illinois,* 2002, pp. 147–152.

42. Adams, M., "PCI Producer Members Weigh SCC Options," *Concrete Products,* October, 2008, pp. PC10–PC11.

43. Sompura, K., "SCC Value in Precast Applications," Presentation at the Spring 2011 American Concrete Institute Convention, Tampa, Florida, April 5, 2011.

44. De Schutter, G., Bartos, P.J.M., Domone, P., Gibbs, J., *Self-Compacting Concrete*, Whittles Publishing, 2008.
45. Daczko, J.A., Vachon, M., "Self-Consolidating Concrete (SCC)," *Significance of Tests and Properties of Concrete & Concrete-Making Materials*, STP 169D, ASTM, 2006, p. 640.
46. Schlagbaum, T., "Economic Impact of Self-Consolidating Concrete (SCC) in Ready-Mixed Concrete," *Proceedings of the First North American Conference on the Design and Use of Self-Consolidating Concrete, Chicago, Illinois*, 2002, pp. 131–136.
47. Sullivan, S., "SCC Value Estimator—A New Tool for Establishing the Value of SCC for the Contractor," Presentation at the Spring, 2011 American Concrete Institute Convention, Tampa, Florida, April 5, 2011.
48. Benedict, D.E., Worsfold, S.J., "Economically Attractive Field Applications for Self-Consolidating Concrete," *Proceedings of the Second North American Conference on the Design and Use of Self-Consolidating Concrete and the Fourth International RILEM Symposium on Self-Compacting Concrete, Chicago, Illinois*, 2005, pp. 1055–1060.
49. Emborg, M., "How to Increase the Market of SCC Experiences from Nordic Countries on RMC," *Proceedings of the Third North American Conference on the Design and Use of Self-Consolidating Concrete, Chicago, Illinois*, 2008.
50. Keller, L., "The Value of SCC in Construction Projects from a Contractor's Perspective," Presentation at the Spring 2011 American Concrete Institute Convention, Tampa, Florida, April 5, 2011.
51. Grauers, M., "Self-Compacting Concrete—The Way to Cost Effective Production of Buildings," *Proceedings of the Second International Symposium on Self-Compacting Concrete, Tokyo, Japan*, 2001, pp. 723–728.
52. Keller, L., Personal communication, June 2011.
53. Okamura, H., Ouchi, M., "Self-Compacting Concrete, Development, Present Use and Future," *Proceedings of the First International Symposium on Self-Compacting Concrete, Stockholm, Sweden*, 1999, p. 4.
54. Billberg, P., Petersson, O., Osterberg, T., " Full Scale Casting of Bridges with Self-Compacting Concrete," *Proceedings of the First International RILEM Symposium on Self-Compacting Concrete, Stockholm, Sweden*, 1999, pp. 639–650.

Chapter 3 Fresh Properties and Test Methods

1. "Terminology," American Concrete Institute Website, www.concrete.org, March 4, 2011.
2. Tattersall, G.H., Banfill, P.F.G., *The Rheology of Fresh Concrete*, Pitman, Boston, 1983.
3. Sedran, T, de Larrard, F., "Optimization of Self-Compacting Concrete Thanks to Packing Model," *Proceedings of the First International Symposium on Self-Compacting Concrete, Stockholm, Sweden*, 1999, pp. 321–332.
4. Hackley, V.A., Ferraris, C., *Guide to Rheological Nomenclature: Measurements in Ceramic Particulate Systems*, NIST Special Publication 946, National Institute of Standards and Technology, 2001.
5. Ferraris, C.L., Brower, L.E., "Comparison of Concrete Rheometers: International Tests at LCPC (Nantes, France) in October, 2000," National Institute for Standards and Technology Interagency Report 6819, September 2001.
6. Ferraris, C.L., Brower, L.E., "Comparison of Concrete Rheometers: International Tests at MB (Cleveland OH, USA) in May, 2003," National Institute for Standards and Technology Interagency Report 7154, September 2004.

7. Wallevik, O., "Rheology—A Scientific Approach to Develop Self-Compacting Concrete," *Proceedings of the Third International Symposium on Self-Compacting Concrete, Reykjavik, Iceland*, August 17–20, 2003, pp. 23–31.
8. ACI 238.1R-08, "Report on Measurements of Workability and Rheology of Fresh Concrete," American Concrete Institute, 2008.
9. Nielsson, I., Wallevik, O., "Rheological Evaluation of Some Empirical Test Methods, Preliminary Results," *Proceedings of the Third International Symposium on Self-Compacting Concrete, Reykjavik, Iceland*, August 17–20, 2003, pp. 59–68.
10. "ACI 237R-07, Self-Consolidating Concrete," *ACI Manual of Concrete Practice*, American Concrete Institute, 2007, pp. 24–26.
11. De Schutter, G., Bartos, P.J.M., Domone, P., Gibbs, J., *Self-Compacting Concrete*, Whittles Publishing, 2008.
12. Koehler, E.P., Fowler, D.W., "Summary of Concrete Workability Test Methods," ICAR Report 105-1, August 2003.
13. "ASTM C 1611/C 1611M—Standard Test Method for Slump Flow of Self-Consolidating Concrete," *Annual Book of ASTM Standards*, Vol. 04.02, 2009, pp. 850–855.
14. "ASTM C 1621/C 1621M—Standard Test Method for Passing Ability of Self-Consolidating Concrete by J-Ring," *Annual Book of ASTM Standards*, Vol. 04.02, 2009, pp. 856–859.
15. *The European Guidelines for Self-Compacting Concrete: Specification, Production and Use*, The European Project Group, 2005.
16. "Recommendation for Construction of Self-Compacting Concrete," Japan Society of Civil Engineers, 1998.
17. Shen, L., Struble, L., Lange, D., "Modeling Dynamic Segregation of Self-Consolidating Concrete," *ACI Materials Journal*, Vol. 106, No. 4, July/August, 2009, pp. 375–380.
18. "ASTM C 1610/C 1610M—Standard Test Method for Static Segregation of Self-Consolidating Concrete Using Column Technique," *Annual Book of ASTM Standards*, Vol. 04.02, 2009, pp. 846–849.
19. "ASTM C 1712/C 1712M—Standard Test Method for Rapid Assessment of Static Segregation Resistance of Self-Consolidating Concrete Using Penetration Test," *Annual Book of ASTM Standards*, Vol. 04.02, 2009.
20. Daczko, J.A., Kurtz, M.A., "Development of High Volume Coarse Aggregate Self-Compacting Concrete," *Proceedings of the Second International Symposium on Self-Compacting Concrete, Tokyo, Japan*, 2001, pp. 403–412.
21. Daczko, J.A., Phillips, S., "Self-Compacting Concrete in Underground and Mining Applications," *Proceedings of the Second International Symposium on Self-Compacting Concrete, Tokyo, Japan*, 2001, pp. 671–680.
22. "ASTM C 232/C 232M—Standard Test Methods for Bleeding of Concrete," *Annual Book of ASTM Standards*, Vol. 04.02, 2009, pp. 156–170.
23. Utsi, S., Emborg, M., Carlsward, J., "Relation between Workability and Rheological Parameters," *Proceedings of the Third International Symposium on Self-Compacting Concrete, Reykjavik, Iceland*, 2003, pp. 154–164.
24. BASF Internal Development Data, Mixtures 2918–2925.
25. Daczko, J.A., "A Comparison of Passing Ability Test Methods for Self-Consolidating Concrete," *Proceedings of the Third International Symposium on Self-Compacting Concrete, Reykjavik, Iceland*, 2003, pp. 335–344.

Chapter 4 Hardened Properties of SCC

1. Konig, G., Holschemacher, K., Dehn, F., Weibe, D., "Self-Compacting Concrete—Time Development of Material Properties and Bond Behaviour,"

Proceedings of the Second International Symposium on Self-Compacting Concrete, Tokyo, Japan, 2001, p. 507.

2. Roziere, E., Turcry, P., Loukili, A., "Influence of Paste Volume, Addition Content and Addition Type on Shrinkage Cracking of Self-Compacting Concrete," *Proceedings of the Second North American Conference on the Design and Use of Self-Consolidating Concrete and the Fourth International RILEM Symposium on Self-Compacting Concrete, Chicago, Illinois,* 2005, pp. 945–951.

3. Hwang, S., Khayat, K.H., "Performance of Hardened Self-Consolidating Concrete Designated for Repair Applications," *Proceedings of the Second North American Conference on the Design and Use of Self-Consolidating Concrete and the Fourth International RILEM Symposium on Self-Compacting Concrete, Chicago, Illinois,* 2005, pp. 965–971.

4. D'Ambrosia, M.D., Lange, D.A., Brinks, A.J., "Restrained Shrinkage and Creep of Self-Consolidating Concrete," *Proceedings of the Second North American Conference on the Design and Use of Self-Consolidating Concrete and the Fourth International RILEM Symposium on Self-Compacting Concrete, Chicago, Illinois,* 2005, pp. 921–928.

5. Tam, C.T., Sheinn, S., Swaddiwudhipong, S., Teng, S.H.N., "Comparing the Volume Change Between SCC and Conventional Slump Concrete," *Proceedings of the Second North American Conference on the Design and Use of Self-Consolidating Concrete and the Fourth International RILEM Symposium on Self-Compacting Concrete, Chicago, Illinois,* 2005, pp. 997–1001.

6. Pineaud, A., Cabrillac, R., Remond, S., Pimienta, P., Rivillon, P., "Mechanical Properties of Self-Compacting Concrete: Influence of Composition Parameters," *Proceedings of the Second North American Conference on the Design and Use of Self-Consolidating Concrete and the Fourth International RILEM Symposium on Self-Compacting Concrete, Chicago, Illinois,* 2005, pp. 863–868.

7. Klug, Y., Holshemacher, K., "Comparison of the Hardened Properties of Self-Compacting and Normal Vibrated Concrete," *Proceedings of the Third International RILEM Symposium of Self-Compacting Concrete, Reykjavik, Iceland,* 2003, p. 596.

8. *European Guidelines for Self-Compacting Concrete: Specification, Production and Use,* European Project Group, 2005.

9. "ACI 237R-07, Self-Consolidating Concrete," *ACI Manual of Concrete Practice,* American Concrete Institute, 2007, pp. 12–14.

10. Khayat, K., Mitchell, D., "Attachment D, Research Description and Findings to Self-Consolidating Concrete for Precast, Prestressed Concrete Bridge Elements, Final Report for NCHRP Project 18-12," May 2008.

11. Laumet, P., Czarnecki, A., Nowak, A., "Comparison of the Mechanical Properties of SCC and Ordinary Concrete," *Proceedings of the Second North American Conference on the Design and Use of Self-Consolidating Concrete and the Fourth International RILEM Symposium on Self-Compacting Concrete, Chicago, Illinois,* 2005, pp. 869–875.

12. Adam, T., Ruehl, M., Gruebl, P., "The Effect of Different Binders and Chemical Admixtures on the Early Hardening of High-Strength Self-Consolidating Concrete," *Proceedings of the Second North American Conference on the Design and Use of Self-Consolidating Concrete and the Fourth International RILEM Symposium on Self-Compacting Concrete, Chicago, Illinois,* 2005, pp. 973–979.

13 Collepardi, M., "The New Concrete," *Grafiche Tintoretto,* 2006, p. 265.

14. Neville, A.M., *Properties of Concrete,* 4th ed., Prentice Hall, 1996.

15. BASF Internal Development Data, Mixtures 1726–1749.

16. See, H., Attiogbe, E., "Performance of Self-Consolidating Concrete Under Restrained Shrinkage," *Shrinkage and Creep of Concrete*, ACI SP 227, American Concrete Institute, 2005, pp. 303–316.
17. Hwang, S., Khayat, K., "Effect of Mixture Composition on Restrained Shrinkage Cracking of Self-Consolidating Concrete Used in Repair," *ACI Materials Journal*, September/October 2008, pp. 499–509.
18. Collepardi, M., Borsoi, A., Collepardi, S., Troli, R., "Strength, Shrinkage and Creep of SCC and Flowing Concrete," *Proceedings of the Second North American Conference on the Design and Use of Self-Consolidating Concrete and the Fourth International RILEM Symposium on Self-Compacting Concrete, Chicago, Illinois*, 2005, pp. 911–919.
19. Burgueno, R., Haq, M., "Effect of SCC Mixture Proportioning on Transfer and Development Length of Prestressing Strand," *Self-Consolidating Concrete for Precast Prestressed Applications*, ACI SP 247, American Concrete Institute, 2007, pp. 105–116.
20. Hegger, J., Will, N., Bulte, S., "Bond Strength and Shear Capacity of Prestressed Beams Made with SCC," *Self-Consolidating Concrete for Precast Prestressed Applications*, ACI SP 247, American Concrete Institute, 2007, pp. 123–138.
21. BASF Internal Development Data, Mixtures 471, 477, 497.
22. "ASTM C 1202—Standard Test Method for Electrical Indication of Concrete's Ability to Resist Chloride Ion Penetration," *Annual Book of ASTM Standards*, Vol. C 04.02, 2009, pp. 648–654.
23. BASF Internal Development Data, Mixtures 561, 563, 565.
24. De Schutter, G., "Durability of SCC from Materials to Structures," *Proceedings of the Sixth International RILEM Symposium on Self-Compacting Concrete, Montreal, Canada*, Vol. 2, 2010, pp. 45–54.
25. BASF Internal Development Data, Mixtures 442–457.
26. BASF Internal Development Data, Mixtures 3426–3433.
27. Daczko, J.A., "Stability of Self-Consolidating Concrete, Assumed or Ensured?" *Proceedings of the First North American Conference on the Design and Use of Self-Consolidating Concrete, Chicago, Illinois*, 2002, pp. 223–228.
28. BASF Internal Development Data, Mixtures 2640–2645.
29. BASF Internal Development Data, Mixtures 2691–2693.

Chapter 5 Powders and Aggregates

1. "ASTM C 150—Standard Specification for Portland Cement," *Annual Book of ASTM Standards*, Vol. 04.01, 2009.
2. "ASTM C 1157—Standard Performance Specification for Hydraulic Cement," *Annual Book of ASTM Standards*, Vol. 04.01, 2009.
3. "ASTM C 618—Standard Specification for Coal Fly Ash and Raw or Calcined Natural Pozzolan for Use in Concrete," *Annual Book of ASTM Standards*, Vol. 04.01, 2008.
4. Vikan, H., Justnes, H., Winnefeld, F., "The Importance of Cement Type on Flow Resistance of Cement Paste," *Proceedings of the Second North American Conference on the Design and Use of Self-Consolidating Concrete and the Fourth International RILEM Symposium on Self-Compacting Concrete, Chicago, Illinois*, 2005, pp. 167–172.
5. Haist, M., Muller, H.S., "Optimization of the Pumpability of Self-Compacting Lightweight Concrete," *Proceedings of the Second North American Conference on the Design and Use of Self-Consolidating Concrete and the Fourth International RILEM Symposium on Self-Compacting Concrete, Chicago, Illinois*, 2005, pp. 195–202.

6. Heirman, G., Vandewalle, L., Van Gemert, D., "Influence of Mineral Additions on Rheology and Stability of SCC Pastes," *Proceedings of the Second North American Conference on the Design and Use of Self-Consolidating Concrete and the Fourth International RILEM Symposium on Self Compacting Concrete, Chicago, Illinois*, 2005, pp. 235–242.

7. de Sensale, G.R., Sabalsagaray, B.S., Derregibus, M.T., Cabrera, J., Marziotte, L., Romay, C., "Characterization of Local Fillers," *Proceedings of the Second North American Conference on the Design and Use of Self-Consolidating Concrete and the Fourth International RILEM Symposium on Self-Compacting Concrete, Chicago, Illinois*, 2005, pp. 151–158.

8. Collepardi, M., "The New Concrete," *Grafiche Tintoretto*, 2006, p. 85.

9. Hewlett, P., ed., *Lea's Chemistry of Cement and Concrete*, 4th ed., Butterworth Heinemann, 1988.

10. "ASTM C 187—Standard Test Method for Normal Consistency of Hydraulic Cement," *Annual Book of ASTM Standards*, Vol. 04.01, 2011.

11. "ACI 318—Building Code requirements for Structural Concrete and Commentary," *ACI Manual of Concrete Practice*, Part 3, American Concrete Institute, 2011, pp. 55–62.

12. Kosmatka, S.H., Kekhoff, B., Panarese, W.C., *Design and Control of Concrete Mixtures*, 14th ed., Portland Cement Association, 2002.

13. Basu, P., Saraswati, S., Chowdhury, S., "Effect of Different Fly Ash on Rheology of Mortar for Self-Compacting Concrete," *Proceedings of the Second North American Conference on the Design and Use of Self-Consolidating Concrete and the Fourth International RILEM Symposium on Self-Compacting Concrete, Chicago, Illinois*, 2005, pp. 257–263.

14. Wallevik, O., "Rheology—A Scientific Approach To Develop Self-Compacting Concrete," *Proceedings of the Third International RILEM Symposium on Self-Compacting Concrete, Reykjavik, Iceland*, 2003, pp. 23–31.

15. Billberg, P., Personal communication, 2011.

16. Rached, M., De Moya, M., Fowler, D.W., "Utilizing Aggregate Characteristics to Minimize Cement Content in Portland Cement Concrete," ICAR Report 401, International Center for Aggregates Research, 2009.

17. Ouchi, M., Edamatsu, Y., "A Simple Evaluation Method For Interaction Between Coarse Aggregate and Mortar Particles in Self-Compacting Concrete," *Proceedings of the First International RILEM Symposium on Self-Compacting Concrete, Stockholm, Sweden*, 1999, pp. 121–130.

18. Khayat, K.H., Hu, C., Laye, JM., "Importance of Aggregate Packing Density on Workability of Self-Consolidating Concrete," *Proceedings of the First North American Conference on the Design and Use of Self-Consolidating Concrete, Chicago, Illinois*, 2002, pp. 55–62.

19. Domone, P.L., "Self-Compacting Concrete: An Analysis of 11 Years of Case Studies," *Cement and Concrete Composites*, Vol. 28, 2006, pp. 197–208.

20. Martys, N., Ferraris, C.F., "Simulation of SCC Flow," *Proceedings of the First North American Conference on the Design and Use of Self-Consolidating Concrete, Chicago, Illinois*, 2002, pp. 27–30.

21. Okamura, H., Ouchi, M., "Self-Compacting Concrete—Development, Present Use and Future," *Proceedings of the First International RILEM Symposium on Self-Compacting Concrete, Stockholm, Sweden*, 1999, pp. 3–13.

22. Erdongen, S.T., Fowler, D.W., "Determination of Aggregate Shape Properties Using X-Ray Tomographic Methods and the Effect of Shape on Concrete Rheology," ICAR Report No. ICAR 106-1, International Center for Aggregates Research, 2005.

23. Wall, J., "Design Production and Mechanical Properties of Expanded Slate Lightweight Self-Consolidating Concrete," *Proceedings of the First North American Conference on the Design and Use of Self-Consolidating Concrete, Chicago, Illinois*, 2002, pp. 187–193.
24. BASF Internal Development Data, Mixtures 2766–2777, 2793–2804.
25. Shakhmenko, G., Birsh, J., "Concrete Mix Design and Optimization," *Second International PhD Symposium in Civil Engineering, Budapest, Hungary*, 1998, http://parspbl.com/downloads/ebooks/concrete/Concrete%20Mix%20Design%20and%20Optimization.pdf.
26. Bui, V.K., Montgomery, D., "Mixture Proportioning Method For Self-Compacting High Performance Concrete with Minimum Paste Volume," *Proceedings of the First International RILEM Symposium on Self-Compacting Concrete, Stockholm, Sweden*, 1999, pp. 373–384.
27. Billberg, P., "Mix Design Model for Self-Compacting Concrete," *Proceedings of the First North American Conference on the Design and Use of Self-Consolidating Concrete, Chicago, Illinois*, 2002, pp. 63–68.
28. NIST Virtual Cement and Concrete Testing Laboratory (VCCTL).

Chapter 6 Chemical Admixtures

1. www.concrete.org/technical/CCT/ACI-Terminology.
2. BASF Internal Development Data, Mixtures 2918–2920, 2926–2928, 2930–2932.
3. BASF Internal Development Data, Mixtures 2895, 2904, 2908.
4. "ASTM C 1610—Standard Test Method for Static Segregation of Self-Consolidating Concrete Using Column Technique," *Annual Book of ASTM Standards*, Vol. 04.02, 2009, pp. 846–849.
5. "ASTM C 232—Standard Test Methods for Bleeding of Concrete," *Annual Book of ASTM Standards*, Vol. 04.02, 2009.
6. "ASTM C 1611 – Standard Test Method for Slump Flow of Self-Consolidating Concrete," *Annual Book of ASTM Standards*, Vol. 04.02, 2009, pp. 850–855.
7. BASF Internal Development Data.
8. BASF Internal Development Data, Cement Interaction Worksheet.
9. Nasvik, J., "The ABCs of SCC: As a Cast-in-Place Contractor, Here's What You Should Know," *Concrete Construction*, March 2003, pp. 40–47.
10. BASF Internal Development Data, VMA Robustness Data.
11. Damone, P.L., "Self-Compacting Concrete: An Analysis of 11 Years of Case Studies," *Cement and Concrete Composites*, Vol. 28, 2006, pp. 197–208.
12. Billberg, P., "Self-Compacting Concrete for Civil Engineering Structures: The Swedish Experience," CBI Report 2:99, Swedish Cement and Concrete Research Institute, 1999.
13. Daczko, J., "The Impact of Workability Retention," *Concrete Products*, October 2009, pp. 21–23.
14. Daczko, J., "Innovative New Admixture for Flexible Slump Flow Retention in SCC Mixtures," *Proceedings of the Sixth International RILEM Symposium on Self-Compacting Concrete and the Fourth North American Conference on the Design and Use of Self-Consolidating Concrete, Montreal, Canada*, 2010.
15. BASF Internal Development Data, Mixtures ZMB 181–184.

Chapter 7 Considerations Prior to Mixture Proportioning

1. Domone, P.L., "Self-Compacting Concrete: An Analysis of 11 Years of Case Studies," *Cement and Concrete Composites*, Vol. 28, 2006, pp. 197–208.

2. Wallevik, O.H., "Rheology—A Scientific Approach to Develop Self-Compacting Concrete," *Proceedings of the Third International RILEM Symposium on Self-Compacting Concrete, Reykjavik, Iceland,* 2003, pp. 23–31.
3. Mizobuchi, T., Yanai, S., Takada, K., Sakata, N., Nobuta, Y., "Field Applications of Self-Compacting Concrete with Advantageous Performances," *Proceedings of the First International RILEM Symposium on Self-Compacting Concrete, Stockholm, Sweden,* 1999, pp. 605–616.
4. Nishizaki, T., Kamada, F., Chikamatsu, R., Kawashima, H., "Application of High Strength Self-Compacting Concrete to Prestressed Concrete Outer Tank for LNG Storage," *Proceedings of the First International RILEM Symposium on Self-Compacting Concrete, Stockholm, Sweden,* 1999, pp. 629–638.
5. Billberg, P., Petersson, O., Osterberg, T., "Full Scale of Casting of Bridges with Self-Compacting Concrete," *Proceedings of the First International RILEM Symposium on Self-Compacting Concrete, Stockholm, Sweden,* 1999, pp. 639–650.
6. Chikamatsu, R., Shinkai, C., Kushigemachi, H., "Application of Low Shrinkage Type Self-Compacting Concrete to an Advanced Large Water Purification Plant," *Proceedings of the First International RILEM Symposium on Self-Compacting Concrete, Stockholm, Sweden,* 1999, pp. 659–668.
7. Pichler, R., "The Use of SCC for Building the Millennium Tower in Vienna," *Proceedings of the First International RILEM Symposium on Self-Compacting Concrete, Stockholm, Sweden,* 1999, pp. 729–731.
8. "ACI 237R-07, Self-Consolidating Concrete," *ACI Manual of Concrete Practice,* American Concrete Institute, 2007, pp. 4–8, 18.
9. Inoue, H., Takeichi, Y., Ohtom, T., "Construction of Rigid Foundation of Underground Diaphragm Walls with Highly Congested Reinforcing Bar Arrangements by Using Self-Compacting Concrete," *Proceedings of the Second International RILEM Symposium on Self-Compacting Concrete, Tokyo, Japan,* 2001, pp. 643–650.
10. Tanaka, M., Mori, K., Shindoh, T., Sakamoto, J., "Application of Self-Compacting Concrete to Steel Segments of Multi-Micro Shield Tunneling Method," *Proceedings of the Second International RILEM Symposium on Self-Compacting Concrete, Tokyo, Japan,* 2001, pp. 651–660.
11. Ohtomo, T., Asaka, S., Kim, J., Park, C., Beak, S., Jung, C., Tsutsumi, Y., "Self-Compacting Concrete Used for Underground Diaphragm Walls of the World's Largest 200,000 kl In-Ground Tanks at Inchon LNG Terminal in Korea," *Proceedings of the Second International RILEM Symposium on Self-Compacting Concrete, Tokyo, Japan,* 2001, pp. 661–670.
12. Johansen, K., Kyltveit, B., "SCC in a Rock Repository for Radioactive Waste," *Proceedings of the Second International RILEM Symposium on Self-Compacting Concrete, Tokyo, Japan,* 2001, pp. 681–686.
13. Walraven, J., "Self-Compacting Concrete in the Netherlands," *Proceedings of the First North American Conference on the Design and Use of Self-Consolidating Concrete, Chicago, Illinois,* 2002, pp. 355–360.
14. Lessard, M., Talbot, C., Phelan, W.S., Baker, D., "Self-Consolidating Concrete Solves Challenging Placement Problems at the Pearson International Airport in Toronto, Canada," *Proceedings of the First North American Conference on the Design and Use of Self-Consolidating Concrete, Chicago, Illinois,* 2002, pp. 367–370.
15. Faornasier, G., Giovambattista, P., Zitzer, L., "Self-Consolidating Concrete in Argentina: Development Program and Applications," *Proceedings of the First North American Conference on the Design and Use of Self-Consolidating Concrete, Chicago, Illinois,* 2002, pp. 389–394.

16. Ramsburg, P., Ludirdja, D., Masik, O., "Utilizing SCC in Double Cycling Production of Double Tee Beams," *Proceedings of the Second North American Conference on the Design and Use of Self-Consolidating Concrete and the Fourth International RILEM Symposium on Self-Compacting Concrete, Chicago, Illinois*, 2005, pp. 1049–1054.

17. Chai, P.H.-W., Yang, M., "Case Study: RC Building Rehabilitated with Self-Compacting Concrete," *Proceedings of the Second North American Conference on the Design and Use of Self-Consolidating Concrete and the Fourth International RILEM Symposium on Self-Compacting Concrete, Chicago, Illinois*, 2005, pp. 1069–1074.

18. Fernandez, P., Luciano, J., Constantiner, D., "Successful Implementation of SCC in a Precast Operation—A Case Study," *Proceedings of the Second North American Conference on the Design and Use of Self-Consolidating Concrete and the Fourth International RILEM Symposium on Self-Compacting Concrete, Chicago, Illinois*, 2005, pp. 1075–1082.

19. Jooste, P., Fanourakis, G., "SCC: The South African Experience," *Proceedings of The Second North American Conference on the Design and Use of Self-Consolidating Concrete and the Fourth International RILEM Symposium on Self-Compacting Concrete, Chicago, Illinois*, 2005, pp. 1119–1124.

20. Vieira, M., Bettencourt, A., "Repair Applications of Self-Compacting Concrete," *Proceedings of the Second North American Conference on the Design and Use of Self-Consolidating Concrete and the Fourth International RILEM Symposium on Self-Compacting Concrete, Chicago, Illinois*, 2005, pp. 1125–1132.

21. Khrapko, M., "Experience with Supplying SCC," *Proceedings of the Second North American Conference on the Design and Use of Self-Consolidating Concrete and the Fourth International RILEM Symposium on Self-Compacting Concrete, Chicago, Illinois*, 2005, pp. 1189–1194.

22. Collepardi, M., Collepardi, S., Ogoumah Olagot, J.J., Troli, R., "Laboratory Tests and Field Experiences of High Performance SCCs," *Proceedings of the Third International RILEM Symposium on Self-Compacting Concrete, Reykjavik, Iceland*, 2003, pp. 904–912.

23. Khayat, K.H., Morin, R., "Performance of Self-Consolidating Concrete used to Repair Parapet Wall in Montreal," *Proceedings of the Third International RILEM Symposium on Self-Compacting Concrete, Reykjavik, Iceland*, 2003, pp. 913–919.

24. Sciaraffia, R., "SCC Application in a Tunnel Lining (Chile)," *Proceedings of the Third International RILEM Symposium on Self-Compacting Concrete, Reykjavik, Iceland*, 2003, pp. 986–994.

25. Utsi, S., Jonasson, J., Wallin, K., Ekman, T., "Use of SCC in a Tunnel Lining for a Railway Tunnel in Sweden," *Proceedings of the Third International RILEM Symposium on Self-Compacting Concrete, Reykjavik, Iceland*, 2003, pp. 995–1004.

26. Paris, M., Tavano, S., Albiero, E., "Self-Compacting Concrete (SCC): The Construction of the Base for the Mill of the Unimed Cement Plant of Barletta (Italy)," *Proceedings of the Third International RILEM Symposium on Self-Compacting Concrete, Reykjavik, Iceland*, 2003, pp. 1017–1025.

27. Titherington, M., Schell, H., "Use of Self-Consolidating Concrete for Bridge Repairs," *Proceedings of the Third North American Conference on the Design and Use of Self-Consolidating Concrete, Chicago, Illinois*, 2008.

28. Talbot, C., "Behavior of Self-Consolidating Concrete at Cryogenic Temperatures," *Proceedings of the Third North American Conference on the Design and Use of Self-Consolidating Concrete, Chicago, Illinois*, 2008.

29. Peterson, E., "Meeting the Formwork Requirements for a Challenging Structure," *Concrete International*, Vol. 32. No. 1, 2010, pp. 42–48.
30. "Interim Guidelines for the Use of Self-Consolidating Concrete in Precast/ Prestressed Concrete Institute Member Plants," TR-6-03, Precast/Prestressed Concrete Institute, 2003.
31. Rodriguez Camacho, R.E., Afif, U., Martinez Corona, G., Montano Ramon, H., Martinez Sanchez, B., "Applications of SCC Technology for Precast/Prestressed Elements in Mexico," *Concrete Plant International*, June 2007, pp. 62–70.
32. Daczko, J., "North American Acceptance of Self-Consolidating: A Diffusion of Innovations Perspective," *Concrete Plant International*, April 2009, pp. 18–21.
33. Thrane, L.N., Pade, C., Nielsen, C., Glavind, M., Schou, J., Idzerda, C., "Self-Consolidating Concrete: In the Low Slump Flow Range, Danish Experiences," Presented at the ACI Fall 2009 Convention, New Orleans, Louisiana.
34. Okamura, H., Ouchi, M., "Self-Compacting Concrete: Development, Present Use and Future," *Proceedings of the First International RILEM Symposium on Self-Compacting Concrete, Stockholm, Sweden*, 1999, pp. 3–13.
35. Tomosawa, F., Masuda, Y., Izumi, I., Hayakawa, M., "AIJ Recommended Practice for High Fluidity Concrete for Building Construction," *Proceedings of the First International RILEM Symposium on Self-Compacting Concrete, Stockholm, Sweden*, 1999, pp. 775–786.
36. Nagataki, S., Kawai, T., Fujiwara, H., "State of the Art Report on SCC in Japan," *Proceedings of SCC2010, Design, Production and Placement of Self-Consolidating Concrete, Montreal, Canada*, 2010, Vol. II, pp. 5–24.
37. *European Guidelines for Self-Compacting Concrete: Specification, Production and Use*, European Project Group, 2005.
38. BASF Internal Report AD-026-99, 1999.
39. Daczko, J.A., Constantiner, D., "Rheodynamic Concrete," *Proceedings of the 43rd Congreso Brasileiro do Concreto*, Instituto Brasileiro do Concreto, 2001.
40. "Interim Guidelines for the Use of Self-Consolidating Concrete in Precast/ Prestressed Concrete Institute Member Plants," TR-6-03, Precast/Prestressed Concrete Institute, 2003.
41. "ACI 309.2R-98, Identification and Control of Visible Effects of Consolidation on Formed Concrete Surfaces," *ACI Manual of Concrete Practice*, American Concrete Institute, 2011, p. 2.
42. Shen, L., Struble, L., Lange, D., "Modeling Dynamic Segregation of Self-Consolidating Concrete," *ACI Materials Journal*, Vol. 106, July/August 2009, pp. 375–380.

Chapter 8 Proportioning SCC Mixtures

1. Kheder, G.F., Al Jadiri, R.S., "New Method for Proportioning Self-Consolidating Concrete Based on Compressive Strength Requirements," *ACI Materials Journal*, Vol. 107, No. 5, September/October 2010, pp. 490–497.
2. Bethmont, S., D'aloia Schwarzentruber, L., Stefani, C., Leroy, R., "Defining the Stability Criterion of a Sphere Suspended in a Cement Paste: A Way to Study the Segregation Risk in Self-Compacting Concrete (SCC)," *Proceedings of the Third International RILEM Symposium on Self-Compacting Concrete, Reykjavik, Iceland*, 2003, pp. 94–105.
3. "Effects of Ingredients on Rheology" and "Effects of Ingredients on Segregation," Robustness of SCC, www.selfconsolidatingconcrete.org/advanced/robustness.html.
4. "ACI Terminology." www.aci-int.org.

5. "ASTM C 33-11—Standard Specification for Concrete Aggregates," *Annual Book of ASTM Standards*, Vol. 04.02, 2009, pp. 12–22.
6. "ACI 211—Standard Practice for Selecting Proportions for Concrete," *ACI Manual of Concrete Practice*, American Concrete Institute, 2011, p. 7.
7. Obla, K., Kim, H., Lobo, C., "Effect of Continuous (Well-Graded) Combined Aggregate Grading on Concrete Performance, Phase A: Aggregate Voids Content (Packing Density)," Project D 340, National Ready-Mixed Concrete Association, May 2007.
8. Koehler, E.P., Fowler, D.W., "ICAR Mixture Proportioning Procedure for Self-Consolidating Concrete," Research Report ICAR 108-1, ICAR Project 108: Aggregates in Self-Consolidating Concrete, International Center for Aggregates Research, 2007.
9. Bui, V.K., Montgomery, D., "Mixture Proportioning Method for Self-Compacting High Performance Concrete with Minimum Paste Volume," *Proceedings of the First RILEM International Symposium on Self-Compacting Concrete, Stockholm, Sweden*, 1999, pp. 373–384.
10. Fuller, W.B., Thompson, S.E., "The Laws of Proportioning Concrete," *Transactions, American Society of Civil Engineers*, Vol. 59, 1907, p. 67.
11. Shekhmenko, G., Birsh, J., "Concrete Mix Design and Optimization," *Proceedings of the Second International PhD Symposium in Civil Engineering, Budapest*, 1998, http://parspbl.com/downloads/ebooks/concrete/Concrete%20Mix%20Design%20and%20Optimization.pdf.
12. Erdogan, S.T., Fowler, D.F., "Determination of Aggregate Shape Properties Using X-Ray Tomographic Methods and the Effect of Shape on Concrete Rheology," ICAR Report No. ICAR 106-1, International Center for Aggregates Research, 2005.
13. Oh, S.G., Noguchi, T., Tomosawa, F., "Toward Mix Design for Rheology of Self-Compacting Concrete," *Proceedings of the First RILEM International Symposium on Self-Compacting Concrete, Stockholm, Sweden*, 1999, pp. 361–372.
14. Pedersen, B., Smeplass, S., "The Relationship Between the Rheological Properties of SCC and the Corresponding Matrix Phase," *Proceedings of the Third International RILEM Symposium on Self-Compacting Concrete, Reykjavik, Iceland*, 2003, pp. 106–116.
15. Billberg, P., "Self-Compacting Concrete for Civil Engineering Structures—The Swedish Experience," CBI Report 2.99, Swedish Cement and Concrete Research Institute, 1999.
16. Okamura, K., Ozawa, H., "Mix Design for Self-Compacting Concrete," *Concrete Library of JSCE*, Vol. 25, 1995, pp. 107–120.
17. Billberg, P., "Fine Mortar Rheology in Mix Design of SCC," *Proceedings of the First RILEM International Symposium on Self-Compacting Concrete, Stockholm, Sweden*, 1999, pp. 47–58.
18. Sheinn, A.M.M., Ho, D.W.S., Tam, C.T., "Effect of Particle Shape on Paste Rheology of SCC," *Proceedings of the Third International RILEM Symposium on Self-Compacting Concrete, Reykjavik, Iceland*, 2003, pp. 232–239.
19. de Sensale, G.R., Sabalsagaray, B.S., Derrigibus, M.T., Cabrera, J., Marziotte, L., Romay, C., "Characterization of Local Fillers," *Proceedings of the Second North American Conference on the Design and Use of Self-Consolidating Concrete and the Fourth International RILEM Symposium on Self-Compacting Concrete, Chicago, Illinois*, 2005, pp. 151–158.
20. "ACI 232.2 R-03—Use of Fly Ash in Concrete," *ACI Manual of Concrete Practice*, Part 2, American Concrete Institute, 2011, p. 11.
21. "ACI 237R-07—Self-Consolidating Concrete," *ACI Manual of Concrete Practice*, American Concrete Institute, 2007, p. 16.

22. "ACI 233R-03—Slag Cement in Concrete and Mortar," *ACI Manual of Concrete Practice*, Part 2, American Concrete Institute, 2011, pp. 7–8.

23. "ACI 234R-06—Guide for the Use of Silica Fume in Concrete," *ACI Manual of Concrete Practice*, Part 2, American Concrete Institute, 2011, p. 17.

24. Wallevik, O., "Rheology—A Scientific Approach to Develop Self-Compacting Concrete," *Proceedings of the Third International RILEM Symposium on Self-Compacting Concrete, Reykjavik, Iceland*, 2003, pp. 23–31.

25. Vikan, H., Justnes, H., "Influence of Silica Fume on Rheology of Cement Paste," *Proceedings of the Third International RILEM Symposium on Self-Compacting Concrete, Reykjavik, Iceland*, 2003, pp. 190–201.

26. Su, N., Hsu, K.C., Chai, H.W., "A Simple Mix Design Method for Self-Compacting Concrete," *Concrete and Concrete Research*, Vol. 31, No. 12, 2001, pp. 1799–1807.

27. Mizobuchi, T., Yanai, S., Takada, K., Sakata, N., Nobuta, Y., "Field Applications of Self-Compacting Concrete with Advantageous Performances," *Proceedings of the First International RILEM Symposium on Self-Compacting Concrete, Stockholm, Sweden*, 1999, pp. 605–616.

28. Nishizaki, T., Kamada, F., Chikamatsu, R., Kawashima, H., "Application of High Strength Self-Compacting Concrete to Prestressed Concrete Outer Tank for LNG Storage," *Proceedings of the First International RILEM Symposium on Self-Compacting Concrete, Stockholm, Sweden*, 1999, pp. 629–638.

29. Billberg, P., Petersson, O., Osterberg, T., "Full Scale Casting of Bridges with Self-Compacting Concrete," *Proceedings of the First International RILEM Symposium on Self-Compacting Concrete, Stockholm, Sweden*, 1999, pp. 639–650.

30. Chikamatsu, R., Shinkai, C., Kushigemachi, H., "Application of Low Shrinkage Type Self-Compacting Concrete to an Advanced Large Water Purification Plant," *Proceedings of the First International RILEM Symposium on Self-Compacting Concrete, Stockholm, Sweden*, 1999, pp. 659–668.

31. Inoue, H., Takeichi, Y., Ohtom, T., "Construction of Rigid Foundation of Underground Diaphragm Walls with Highly Congested Reinforcing Bar Arrangements by Using Self-Compacting Concrete," *Proceedings of the Second International RILEM Symposium on Self-Compacting Concrete, Tokyo, Japan*, 2001, pp. 643–650.

32. Tanaka, M., Mori, K., Shindoh, T., Sakamoto, J., "Application of Self-Compacting Concrete to Steel Segments of Multi-Micro Shield Tunneling Method," *Proceedings of the Second International RILEM Symposium on Self-Compacting Concrete, Tokyo, Japan*, 2001, pp. 651–660.

33. Ohtomo, T., Asaka, S., Kim, J., Park, C., Beak., S., Jung, C., Tsutsumi, Y., "Self-Compacting Concrete Used for Underground Diaphragm Walls of the World's Largest 200,000 kl In-Ground Tanks at Inchon LNG Terminal in Korea," *Proceedings of the Second International RILEM Symposium on Self-Compacting Concrete, Tokyo, Japan*, 2001, pp. 661–670.

34. Johansen, K., Kyltveit, B., "SCC in a Rock Repository for Radioactive Waste," *Proceedings of the Second International RILEM Symposium on Self-Compacting Concrete, Tokyo, Japan*, 2001, pp. 681–686.

35. Lessard, M., Talbot, C., Phelan, W.S., Baker, D., "Self-Consolidating Concrete Solves Challenging Placement Problems at the Pearson International Airport in Toronto, Canada," *Proceedings of the First North American Conference on the Design and Use of Self-Consolidating Concrete, Chicago, Illinois*, 2002, pp. 367–370.

36. Faornasier, G., Giovambattista, P., Zitzer, L., "Self-Consolidating Concrete in Argentina: Development Program and Applications," *Proceedings of the First*

North American Conference on the Design and Use of Self-Consolidating Concrete, Chicago, Illinois, 2002, pp. 389–394.

37. Ramsburg, P., Ludirdja, D., Masik, O., "Utilizing SCC in Double Cycling Production of Double Tee Beams," *Proceedings of the Second North American Conference on the Design and Use of Self-Consolidating Concrete and the Fourth International RILEM Symposium on Self-Compacting Concrete, Chicago, Illinois*, 2005, pp. 1049–1054.

38. Jooste, P., Fanourakis, G., "SCC: The South African Experience," *Proceedings of the Second North American Conference on the Design and Use of Self-Consolidating Concrete and the Fourth International RILEM Symposium on Self-Compacting Concrete, Chicago, Illinois*, 2005, pp. 1119–1124.

39. Vieira M., Bettencourt, A., "Repair Applications of Self-Compacting Concrete," *Proceedings of the Second North American Conference on the Design and Use of Self-Consolidating Concrete and the Fourth International RILEM Symposium on Self-Compacting Concrete, Chicago, Illinois*, 2005, pp. 1125–1132.

40. Khrapko, M., "Experience with Supplying SCC," *Proceedings of the Second North American Conference on the Design and Use of Self-Consolidating Concrete and the Fourth International RILEM Symposium on Self-Compacting Concrete, Chicago, Illinois*, 2005, pp. 1189–1194.

41. Collepardi, M., Collepardi, S., Ogoumah Olagot, J.J., Troli, R., "Laboratory Tests and Field Experiences of High Performance SCCs," *Proceedings of the Third International RILEM Symposium on Self-Compacting Concrete, Reykjavik, Iceland*, 2003, pp. 904–912.

42. Khayat, K.H., Morin, R., "Performance of Self-Consolidating Concrete used to Repair Parapet Wall in Montreal," *Proceedings of the Third International RILEM Symposium on Self-Compacting Concrete, Reykjavik, Iceland*, 2003, pp. 913–919.

43. Sciaraffia, R., "SCC Application in a Tunnel Lining (Chile)," *Proceedings of the Third International RILEM Symposium on Self-Compacting Concrete, Reykjavik, Iceland*, 2003, pp. 986–994.

44. Utsi, S., Jonasson, J., Wallin, K., Ekman, T., "Use of SCC in a Tunnel Lining for a Railway Tunnel in Sweden," *Proceedings of the Third International RILEM Symposium on Self-Compacting Concrete, Reykjavik, Iceland*, 2003, pp. 995–1004.

45. Paris, M., Tavano, S., Albiero, E., "Self-Compacting Concrete (SCC): The Construction of the Base for the Mill of the Unimed Cement Plant of Barletta (Italy)," *Proceedings of the Third International RILEM Symposium on Self-Compacting Concrete, Reykjavik, Iceland*, 2003, pp. 1017–1025.

46. Talbot, C., "Behavior of Self-Consolidating Concrete at Cryogenic Temperatures," *Proceedings of the Third North American Conference on the Design and Use of Self-Consolidating Concrete, Chicago, Illinois*, 2008.

47. "Interim Guidelines for the Use of Self-Consolidating Concrete in Precast/Prestressed Concrete Institute Member Plants," TR-6-03, Precast/Prestressed Concrete Institute, 2003.

48. Galeota, D., Giammatteo, M., Gregori, A., Manfredi, G., Marino, R., Vivaldi, S., "SCC for Rapid Home Construction after a Natural Disaster," *Proceedings of SCC2010: Design, Production and Placement of Self-Consolidating Concrete, Montreal, Canada*, 2010, pp. 1369–1377.

49. Domone, P.L., "Self-Compacting Concrete; An Analysis of 11 Years of Case Studies," *Cement and Concrete Composites*, Vol. 28, No. 2, 2006, pp. 197–208.

50. Daczko, J.A., "A Comparison of Passing Ability Test Methods for Self-Consolidating Concrete," *Proceedings of the Third International RILEM*

Symposium on Self-Compacting Concrete, Reykjavik, Iceland, 2003, pp. 335–344.

Chapter 9 Mixture Qualification

1. "Interim Guidelines for the Use of Self-Consolidating Concrete in Precast/ Prestressed Concrete Institute Member Plants," TR-6-03, Precast/Prestressed Concrete Institute, 2003.
2. Anderson, V.L., McLean, R.A., *Design of Experiments: A Realistic Approach,* Marcel Dekker, 1974.
3. Rasch, D., Herrendorfer, G., *Experimental Design, Sample Size Determination and Block Designs,* D. Reidel, 1986.
4. Khayat, K.H., Ghezal, A., Hadriche, M.S., "Utility of Statistical Models in Proportioning Self-Consolidating Concrete," *Proceedings of the First International RILEM Symposium on Self-Compacting Concrete, Stockholm, Sweden,* 1999, pp. 345–360.
5. Fernandez, P., Luciano, J., Constantiner, D., "Successful Implementation of SCC in a Precast Operation—A Case Study," *Proceedings of the Second North American Conference on the Design and Use of Self-Consolidating Concrete and the Fourth International RILEM Symposium on Self-Compacting Concrete, Chicago, Illinois,* 2005, pp. 1075–1082.
6. "ASTM C 1611/C 1611M-05—Standard Test Method for Slump Flow of Self-Consolidating Concrete," *Annual Book of ASTM Standards,* Vol. 04.02, 2009, pp. 850–855.
7. ASTM C 1621/C 1621M-08—Standard Test Method for Passing Ability of Self-Consolidating Concrete by J-Ring," *Annual Book of ASTM Standards,* Vol. 04.02, 2009, pp. 856–859.
8. *European Guidelines for Self-Compacting Concrete: Specification, Production and Use,* European Project Group, May 2005.
9. "Self-Compacting Concrete: Recommendations for Use," Concrete Report No. 10(E), Swedish Concrete Association, 2002.
10. "ASTM C 232/C 232M—Standard Test Methods for Bleeding of Concrete," *Annual Book of ASTM Standards,* Vol. 04.02, 2009, pp. 166–170.
11. "ASTM C 1610/C 1610M-06a—Standard Test Method for Static Segregation of Self-Consolidating Concrete Using Column Technique," *Annual Book of ASTM Standards,* Vol. 04.02, 2009, pp. 846–849.
12. Lange, D.A., Struble, L., D'Ambrosia, M., Birch, B., Shen, L., Tejeda, F., "Self-Consolidating Concrete: The Good the Bad and the Ugly," PowerPoint presentation, http://netfiles.uiuc.edu/dlange/www/research/presentations/LSU goodbadugly.ppt.
13. "ASTM C 231—Standard Test Method for Air Content of Freshly Mixed Concrete by the Pressure Method," *Annual Book of ASTM Standards,* Vol. 04.02, 2009, pp. 156–165.
14. "ASTM C 94/C 94M 07—Standard Specification for Ready-Mixed Concrete," *Annual Book of ASTM Standards,* Vol. 04.02, 2009, pp. 54–64.
15. Rigueira, J.W., Garcia-Taengua, E., Serna-Ros, P., "Self-Consolidating Concrete Robustness in Continuous Production Regarding Fresh and Hardened Properties," *ACI Materials Journal,* Vol. 106, No.3, 2009, pp. 301–307.
16. Domone, P.L., "Self-Compacting Concrete: An Analysis of 11 Years of Case Studies," *Cement and Concrete Composites,* Vol. 28, 2006, pp. 197–208.
17. Bonen, D., Deshpande, Y., Olek, J., Shen, L., Struble, L., Lange, D., Khayat, K., "Robustness of SCC," www.selfconsolidating concrete.org.

18. Billberg, P., "Self-Compacting Concrete for Civil Engineering Structures—The Swedish Experience," CBI Report 2.99, Swedish Cement and Concrete Research Institute, 1999.

Chapter 10 Concrete Batching and Mixing

1. Ouchi, M., Nakjima, Y., "A Guide for Manufacturing and Construction of Self-Compacting Concrete—Learning from Real Troubles," Handout at the Second International RILEM Symposium on Self-Compacting Concrete, Tokyo, 2001.
2. "ACI 304R-00—Guide for Measuring, Mixing, Transporting and Placing Concrete," ACI Manual of Concrete Practice, Part 2, American Concrete Institute, 2010, pp. 3–14.
3. MNL 116—Manual for Quality Control for Plants and Production of Structural Precast, Precast/Prestressed Concrete Institute (PCI), 1999.
4. McCraven, S., "Working with SCC Needn't Be Hit or Miss," Precast Inc. Magazine, July/August, 2010, http://precast.org/2010/07/working-with-scc-neednt-be-hit-or-miss.
5. de Larrard, F., Cazacliu, B., Chopin, D., Chateau, E., "Production of SCC," Proceedings of the Third International Symposium on Self-Compacting Concrete, Reykjavik, Iceland, 2003, pp. 481–494.
6. National Ready-Mixed Concrete Association, "2010 NRMCA Quality Benchmarking Survey," Concrete In Focus Magazine, January/February 2010, pp. 11–15.
7. European Guidelines for Self-Compacting Concrete: Specification, Production and Use, European Project Group, 2005.
8. "ACI 237R-07—Self-Consolidating Concrete," ACI Manual of Concrete Practice, American Concrete Institute, 2007, p. 12.
9. "ASTM C 566—Test Method for Total Evaporable Moisture Content of Aggregate by Drying," Annual Book of ASTM Standards, Vol. 04.02, 2009, pp. 311–313.
10. "ASTM C 70—Standard Test Method for Surface Moisture in Fine Aggregate," Annual Book of ASTM Standards, Vol. 04.02, 2009, pp. 38–40.
11. "Interim Guidelines for the Use of Self-Consolidating Concrete in Precast/Prestressed Concrete Institute Member Plants," TR-6-03, Precast/Prestressed Concrete Institute, 2003.
12. Cazacliu, B., de Larrard, F., Chopin, D., Shell, R., Chateau, E., "Water Control, Homogenization Control, and Transport: Three Critical Stages to Obtain Consistent SCC," Proceedings of the Second North American Conference on the Design and Use of Self-Consolidating Concrete and the Fourth International Symposium on Self-Compacting Concrete, Chicago, Illinois, 2005, Vol. 2, pp. 783–791.
13. Beitzel, H., "Optimisation of the Mixing Process for Producing Self-Compacting High-Performance Concrete," Proceedings of SCC2010, Design, Production and Placement of Self-Consolidating Concrete, Montreal, Canada, 2010, pp. 175–186.
14. Beitzel, H., "Quality Control by Varying Mechanical and Operational Parameters for the Production and Processing of Self-Compacting Concrete," Proceedings of the Second North American Conference on the Design and Use of Self-Consolidating Concrete and the Fourth International Symposium on Self-Compacting Concrete, Chicago, Illinois, 2005, Vol. 2, pp. 705–711.
15. www.concretemixers.biz, March 19, 2011.
16. Miller, B., "Batching and Mixing Equipment," Precast Inc. Magazine, 2005, http://precast.org/2010/05/batching-and-mixing-equipment.

17. Daczko, J.A., "A Proposal for Measuring the Rheology of Production Concrete," *Concrete International*, May 2000, pp. 47–49.
18. Ferraris, C.F., Amziane, S., Koehler, E., "Concrete Mixing Trucks and Workability," *CPI—Concrete Plant International*, No. 3, June 2006, pp. 192–197.
19. Wallevik, O., "Rheology—A Scientific Approach to Develop Self-Compacting Concrete," *Proceedings of the Third International Symposium on Self-Compacting Concrete, Reykjavik, Iceland*, 2003, pp. 23–31.
20. Corradi, M., Khurana, R., Magarotto, R., "User Friendly Self-Compacting Concrete in Precast Production," *Proceedings of the Third International Symposium on Self-Compacting Concrete, Reykjavik, Iceland*, 2003, pp. 457–466.
21. Yelton, R., "Up Close and Direct, In-Mixer Monitor Provides Rheological Measurements for Tighter Quality Control of SCC," *The Concrete Producer*, February 2010, p. 14.
22. High Performance Concrete Technology Delivery Team, Federal Highway Administration, *High Performance Concrete Structural Designers Guide*, 1st ed., US Department of Transportation, March 2005.

Chapter 11 Quality Control of SCC Materials and Mixtures

1. "ACI Terminology," www.aci-int.org, March 20, 2011.
2. National Ready-Mixed Concrete Association, "2010 NRMCA Quality Benchmarking Survey," *Concrete In Focus Magazine*, January/February 2011, pp. 11–15.
3. *MNL 116—Manual for Quality Control for Plants and Production of Structural Precast*, Precast/Prestressed Concrete Institute (PCI), 1999.
4. *Quality Control Manual for Precast and Prestressed Concrete Plants*, 6th ed., National Precast Concrete Association (NPCA), November 2007.
5. "ASTM C 117-04—Standard Test Method for Materials Finer than 75-µm (No. 200) Sieve in Mineral Aggregates by Washing," *Annual Book of ASTM Standards*, Vol. 04.02, 2009, pp. 65–67.
6. Billberg, P., "Self-Compacting Concrete for Civil Engineering Structures—The Swedish Experience," CBI Report 2.99, Swedish Cement and Concrete Research Institute, 1999.
7. *European Guidelines for Self-Compacting Concrete: Specification, Production and Use*, European Project Group, 2005.
8. Nagataki, S., Kawai, T., Fujiwara, H., "State of the Art Report on SCC in Japan," *Proceedings of SCC2010, Design, Production and Placement of Self-Consolidating Concrete, Montreal, Canada*, 2010, Vol. II, pp. 5–24.
9. "ASTM C 1621/C 1621M—Standard Test Method for Passing Ability of Self-Consolidating Concrete by J-Ring," *Annual Book of ASTM Standards*, Vol. 04.02, 2009, pp. 856–859.
10. Kubo, M., Nakano, M., Aoki, H., Sugano, S., Ouchi, M., "The Quality Control Method of Self-Compacting Concrete Using Testing Apparatus for Self-Compactability Evaluation," *Proceedings of the Second International RILEM Symposium on Self-Compacting Concrete, Tokyo, Japan*, 2001, pp. 555–564.
11. "ASTM C 1610/C 1610M—Standard Test Method for Static Segregation of Self-Consolidating Concrete Using Column Technique," *Annual Book of ASTM Standards*, Vol. 04.02, 2009, pp. 846–849.
12. "ASTM C 1712/C 1712M—Standard Test Method for Rapid Assessment of Static Segregation Resistance of Self-Consolidating Concrete Using Penetration Test," *Annual Book of ASTM Standards*, Vol. 04.02.
13. "ASTM C 1611/C 1611M—Standard Test Method for Slump Flow of Self-Consolidating Concrete," *Annual Book of ASTM Standards*, Vol. 04.02, 2009, pp. 850–855.

14. Mizobuchi, T. Yanai, S., Takada, K., Sakata, N., Nobuta, Y., "Field Applications of Self-Compacting Concrete with Advantageous Performances," *Proceedings of the First International RILEM Symposium on Self-Compacting Concrete, Stockholm, Sweden*, 1999, pp. 605–616.

15. Nishizaki, T., Kamada, F., Chikamatsu, R., Kawashima, H., "Application of High Strength Self-Compacting Concrete to Prestressed Concrete Outer Tank for LNG Storage," *Proceedings of the First International RILEM Symposium on Self-Compacting Concrete, Stockholm, Sweden*, 1999, pp. 629–638.

16. Billberg, P., Petersson, O., Osterberg, T., "Full Scale Casting of Bridges with Self-Compacting Concrete," *Proceedings of the First International RILEM Symposium on Self-Compacting Concrete, Stockholm, Sweden*, 1999, pp. 639–650.

17. Chikamatsu, R., Shinkai, C., Kushigemachi, H., "Application of Low Shrinkage Type Self-Compacting Concrete to an Advanced Large Water Purification Plant," *Proceedings of the First International RILEM Symposium on Self-Compacting Concrete, Stockholm, Sweden*, 1999, pp. 659–668.

18. Inoue, H., Takeichi, Y., Ohtom, T., "Construction of Rigid Foundation of Underground Diaphragm Walls with Highly Congested Reinforcing Bar Arrangements by Using Self-Compacting Concrete," *Proceedings of the Second International RILEM Symposium on Self-Compacting Concrete, Tokyo, Japan*, 2001, pp. 643–650.

19. Tanaka, M., Mori, K., Shindoh, T., Sakamoto, J., "Application of Self-Compacting Concrete to Steel Segments of Multi-Micro Shield Tunneling Method," *Proceedings of the Second International RILEM Symposium on Self-Compacting Concrete, Tokyo, Japan*, 2001, pp. 651–660.

20. Ohtomo, T., Asaka, S., Kim, J., Park, C., Beak., S., Jung, C., Tsutsumi, Y., "Self-Compacting Concrete Used for Underground Diaphragm Walls of the World's Largest 200,000 kl In-Ground Tanks at Inchon LNG Terminal in Korea," *Proceedings of the Second International RILEM Symposium on Self-Compacting Concrete, Tokyo, Japan*, 2001, pp. 661–670.

21. Johansen, K., Kyltveit, B., "SCC in a Rock Repository for Radioactive Waste," *Proceedings of the Second International RILEM Symposium on Self-Compacting Concrete, Tokyo, Japan*, 2001, pp. 681–686.

22. Faornasier, G., Giovambattista, P., Zitzer, L., "Self-Consolidating Concrete in Argentina: Development Program and Applications," *Proceedings of the First North American Conference on the Design and Use of Self-Consolidating Concrete, Chicago, Illinois*, 2002, pp. 389–394.

23. Khayat, K.H., Morin, R., "Performance of Self-Consolidating Concrete used to Repair Parapet Wall in Montreal," *Proceedings of the Third International RILEM Symposium on Self-Compacting Concrete, Reykjavik, Iceland*, 2003, pp. 913–919.

24. Sciaraffia, R., "SCC Application in a Tunnel Lining (Chile)," *Proceedings of the Third International RILEM Symposium on Self-Compacting Concrete, Reykjavik, Iceland*, 2003, pp. 986–994.

25. Utsi, S., Jonasson, J., Wallin, K., Ekman, T., "Use of SCC in a Tunnel Lining for a Railway Tunnel in Sweden," *Proceedings of the Third International RILEM Symposium on Self-Compacting Concrete, Reykjavik, Iceland*, 2003, pp. 995–1004.

26. Paris, M., Tavano, S., Albiero, E., "Self-Compacting Concrete (SCC): The Construction of the Base for the Mill of the Unimed Cement Plant of Barletta (Italy)," *Proceedings of the Third International RILEM Symposium on Self-Compacting Concrete, Reykjavik, Iceland*, 2003, pp. 1017–1025.

27. Titherington, M., Schell, H., "Use of Self-Consolidating Concrete for Bridge Repairs," *Proceedings of the Third North American Conference on the Design and Use of Self-Consolidating Concrete, Chicago, Illinois*, 2008.
28. "Recommendation for Construction of Self-Compacting Concrete," Japan Society of Civil Engineers, 1998.
29. "Interim Guidelines for the Use of Self-Consolidating Concrete in Precast/ Prestressed Concrete Institute Member Plants," TR-6-03, Precast/Prestressed Concrete Institute, 2003.
30. de Larrard, F., Cazacliu, B., Chopin, D., Chateau, E., "Production of SCC," *Proceedings of the Third International Symposium on Self-Compacting Concrete, Reykjavik, Iceland*, 2003, pp. 481–494.
31. Cazacliu, B., de Larrard, F., Chopin, D., Shell, R., Chateau, E., "Water Control, Homogenization Control, and Transport: Three Critical Stages to Obtain Consistent SCC," *Proceedings of the Second North American Conference on the Design and Use of Self-Consolidating Concrete and the Fourth International Symposium on Self-Compacting Concrete, Chicago, Illinois*, 2005, Vol. 2, pp. 783–791.

Chapter 12 Transport, Placement, Finishing, and Curing

1. "ACI 304R-00—Guide for Measuring, Mixing, Transporting and Placing Concrete," *ACI Manual of Concrete Practice*, Part 2, American Concrete Institute, 2011, pp. 10–14.
2. "ACI 237R-07—Self-Consolidating Concrete," *ACI Manual of Concrete Practice*, American Concrete Institute, 2007, p. 20.
3. Skarendahl, A., Billberg, P., "Casting of Self-Compacting Concrete," RILEM Report 35 (Final Report of RILEM Committee TC 188 CSC—Casting of Self-Compacting Concrete), 2006.
4. Lange, D.A., Struble, L., D'Ambrosia, M., Birch, B., Shen, L., Tejeda, F., "Self-Consolidating Concrete: The Good the Bad and the Ugly," PowerPoint presentation, http://netfiles.uiuc.edu/dlange/www/research/presentations/LSU goodbadugly.ppt.
5. Khan, L.F., Kurtis, K., "Self-consolidating Concrete in Congested Sections: Mixture Characteristics and Assessment of Performance," *PCI Journal*, Vol. 55, No.1, 2010, pp. 79–96.
6. Vikan, H., "Quality of Concrete Surfaces, State of the Art," SINTEF Report SBF BK A07013, 2007.
7. *European Guidelines for Self-Compacting Concrete: Specification, Production and Use*, European Project Group, 2005.
8. "DIN Standard on Formwork Pressures Updated; Standard Now Addresses Pressures Exerted by Self-Consolidating Concrete," *Concrete International*, June 2010, pp. 27–29.
9. "State of the Art Review of Form Pressure Exerted by Self-Consolidating Concrete," Prepared by the University of Sherbrooke for the RMC Research and Education Foundation in Collaboration with the ACI Strategic Development Council, July 2010.
10. Khayat, K.H., Omran, A.F., D'Amobrosia, M., "Prediction of SCC Formwork Pressure in Full-Scale Elements," *Proceedings of SCC2010, Design, Production and Placement of Self-Consolidating Concrete, Montreal, Canada*, 2010, pp. 231–242.
11. BASF Internal Development Data, Mixtures 514, 515, 545, 546.
12. "Self-Compacting Concrete, Recommendations for Use," Concrete Report No. 10(E), Swedish Concrete Association, 2002.

13. "ACI 308R-01—Guide to Curing Concrete," *ACI Manual of Concrete Practice*, Part 2, American Concrete Institute, 2011, pp. 1–26.

Chapter 13 Troubleshooting SCC Performance

1. Vikan, H., "Quality of Concrete Surfaces, State of the Art," SINTEF Report SBF BK A07013, 2007.
2. Floyd, R.W., Ruiz, E.D., Do, N.H., Hale, W.M., "Development Lengths of High Strength Self-Consolidating Concrete Beams," *PCI Journal*, Vol 56, No.1, 2011, pp. 36–52.
3. Daczko, J.A., Phillips, S.H.E., "Self-Compacting Concrete in Underground and Mining Applications," *Proceedings of the Second International RILEM Symposium on Self-Compacting Concrete, Tokyo, Japan*, 2001, pp. 671–680.
4. "ASTM C 172—Standard Practice for Sampling Freshly Mixed Concrete," *Annual Book of ASTM Standards*, Vol. 04.02, 2009, pp. 122–124.

Chapter 15 Applications of SCC

1. Morrow, R., Amekuedi, G., "Rebuilding a Landmark: Self-Consolidating Concrete Helps Construction Team Win the Battle of Concrete Placement," *Structure Magazine*, July 2005, pp. 31–33.
2. "Concrete at its Best 2005," *Concrete Construction Magazine*, January 2006, p. 37.
3. "Slag Cement Association Project Awards," *Concrete International*, January 2011, pp. 22–23.
4. The Oceans Exhibit, www.irmca.com/projects/2007projects, March 30, 2011.
5. MacDonald, K., "Putting the Hurt Behind," *The Concrete Producer*, December 2008, pp. 40–46.
6. Lwin, M., *HPC Bridge Views*, Issue 50, July/August 2008.
7. Peterson, E., "Meeting the Formwork Requirements for a Challenging Structure," *Concrete International*, January 2010, pp. 42–47.

Index

Printed in the United States
by Baker & Taylor Publisher Services